Defence Works
Functional Standards

Hot Rolled Asphalt and Coated Macadam for … t Works

…fence Works Services
Ministry of Defence

London: HMSO

ISBN 011 772688 5

Acknowledgements

This document has been produced by the Transport Research Laboratory under commission to Defence Works Services, Ministry of Defence. It has been developed in conjunction with the Airfields and Bulk Fuels Group/DWS and has in part evolved from standard specifications previously produced by former Government Departments: Air Ministry Works Department, Ministry of Public Buildings and Works and the Property Services Agency.

Forward

This document is for the use of Top Level Budget Holders (TLBHs) for application by the Project Sponsors and their Project Managers, Property Managers (PROM), Establishment Works Consultants (EWC), Works Service Managers (WSM) and other parties involved with airfield pavement works.

This Defence Works Functional Standard supersedes the following DOE/PSA Specifications for airfield pavement works:

a. PSA Standard Specification Clauses for Airfield Pavement Works Part 4 (1979) - Sections covering Rolled Asphalt Wearing Course.

b. PSA Standard Specification Clauses for Airfield Pavement Works Part 4 (1989) - Sections covering Macadam Base Course and Pervious Macadam.

This Functional Standard was prepared under the patronage of HQ Strike Command for application to airfield pavement works on the MOD estate.

The application and limitations of the specification requirements in this Functional Standard are outlined in Section 1. Further technical assistance regarding the contents of this document can be obtained from DWS. Approaches may be made through local DWS offices or directly to the airfield pavement Technical Works Authority, (DWS TA):

Head of Airfield Pavements
Room 120
Airfields and Bulk Fuels Group
P O Box 1734
Sutton Coldfield
West Midlands
B75 7QB

This DWS Functional Standard, "*Hot Rolled Asphalt and Coated Macadam for Airfield Pavement Works*", has been devised for use of the Crown and of its Contractors in the execution of contracts for the Crown and, subject to the Unfair Contracts Terms Action 1977, the Crown will not be liable in any way whatever (including but without limitation negligence on the part of the Crown its servants or agents) where the Standard is used for other purposes.

Glossary of Technical Terms

Asphaltic Concrete	A mixture of continuous graded aggregate, filler and bituminous binder proportioned to produce a dense and impermeable surfacing.
Basecourse	The layer or layers of the bituminous surfacing immediately below the wearing course.
Binder	A material used for the purpose of holding solid particles together as a coherent mass.
Bitumen Emulsion	An emulsion in which bitumen is dispersed in water or in aqueous solution with the aid of suitable emulsifying agents.
Bituminous	Containing road tar, bitumen, pitch or mixtures thereof.
Bituminous Surfacing	A friction course, wearing course, or a combination of these and a basecourse.
Cutback Bitumen	Petroleum bitumen whose viscosity has been reduced by the addition of a suitable volatile diluent.
Dense Bitumen Macadam	See Macadam.
(Sieved) Fraction	That portion of aggregate which passes one BS sieve but is retained on the adjacent smaller sized sieve in the sequence of sieves used to specify that grading.
Filler	A finely divided mineral powder added to road tar, bitumen or the like, or to a mixture containing the same, in order to effect some desired change in the properties of the binding material.
Friction Course	A relatively thin layer of 10 mm nominal sized open-graded Macadam which allows free penetration of the surface water to the underlying impervious wearing course.
Heavy Duty Macadam (HDM)	See Macadam.
Hot Rolled Asphalt (HRA)	A mixture of gap-graded aggregate, filler and bitumen binder proportioned to a design or recipe to produce a dense and impermeable surfacing material.
Macadam	A graded aggregate coated with tar or bitumen. a. Dense Bitumen Macadam (DBM): A dense, relatively impermeable, Macadam coated with a bitumen binder and with a filler content of between 2 per cent and 9 per cent.

b. Heavy Duty Macadam (HDM): A dense bitumen Macadam with 50 pen bitumen binder and a high filler content of 7 per cent to 11 per cent.

c. Pervious Macadam: A layer of open-graded Macadam with a maximum nominal size aggregate of 28 mm which acts as a topping to protect whilst allowing free penetration of the surface water to French drains.

Marshall Asphalt	An Asphaltic Concrete with graded bitumen bound aggregates designed to achieve specified stability, flow, voids and density characteristics.
Pavement	A structure consisting of a layer or superimposed layers of selected materials, whose primary purpose is to distribute the applied load to the subgrade.
Pervious Macadam	See Macadam.
Petroleum Bitumen	Bitumen obtained from crude oil by refinery processes.
Regulating Material	Bituminous material of variable thickness applied to an existing pavement to adjust the shape preparatory to resurfacing.
Road Tar	A viscous liquid, derived from crude tar obtained by the destructive distillation of coal.
Tack Coat	A thin film of binder to improve the adhesion between two courses.
Wearing course	The layer of the bituminous surfacing immediately below the friction course, or which directly supports the traffic.

NOTE. The Project/Works Services Manager should delete any terms not applicable to his project and should add any terms necessary due to the particular nature of his project.

Contents

Page No.

1 Introduction

1.1 Background

1.1.1 The unique characteristics of military aircraft, in terms of speed, weight, tyre pressures, etc., create specialist requirements in the surfacing of MOD airfields. As such, specialist materials specifications are required to meet these needs.

1.1.2 This Standard, for Hot Rolled Asphalt and Coated Macadam, is one of a series being produced by DWS to lay down specification requirements for airfield pavement works. The following clauses in this Section are intended to set out the applications of Hot Rolled Asphalt and Coated Macadam in the construction and refurbishment of MOD airfield pavements.

1.1.3 The use of this Standard does not absolve a Project/Works Services Manager from any responsibility for his designs, neither does its existence constrain him from using alternatives, provided such alternatives can be demonstrated to provide a result of equal quality.

1.2 Functional Requirements of Airfield Pavements

1.2.1 The pavements must facilitate safe aircraft ground operations. In order to do this they must meet certain specialist performance requirements. The following sets out the main requirements, the relative importance of which will be dependent on the function of the pavements and the nature and type of aircraft operations:

a. Good rideability.

b. Good friction characteristics.

c. High strengths and stability to withstand the shear stresses induced by heavy wheel loads and high tyre pressures.

d. A durable, hard-wearing weatherproof surface free from loose material and sharp edges which might endanger aircraft.

e. Resistance to fuel spillage and jet blast. Depending on the nature and type of aircraft operations, these requirements are likely to be too onerous for bituminous surfacings in certain areas of the airfield.

f. Facilitate economic maintenance.

1.2.2 Main Length of Runway

On the main length of a runway, it is important to have good rideability and wet weather skid resistance. Bituminous surfacings generally give better rideability than concrete, with the latter being partly dependent on the spacing and detail of the transverse joints. Good skid resistance/friction characteristics are achieved by a combination of good surface slope to effect efficient surface water run-off and also surface texture to increase friction and allow water to disperse from under the tyre contact area. The standard technique used in the UK for providing texture on bituminous road surfacings is by the provision of rolled-in chippings. This is not acceptable on airfield pavements because materials which are likely to pluck out or spall on the surface of a pavement are a potential FOD (Foreign Object Damage) hazard to aircraft, particularly with regard to ingestion by jet engines. The methods currently favoured for providing suitable friction characteristics on new bituminous surfacings on MOD runways are as follows:

a. The provision of a 20 mm thick Friction Course (not included in this Standard) on Asphalt surfacing. This is the most favoured construction option but it would not be specified in certain situations including dusty regions, very cold climates, runways used for STOL (short take-offs and landings) operations and where night-time construction work necessitates numerous transverse construction joints.

b. By grooving an Asphalt surface with continuous grooves cut perpendicular to the centre-line of the runway.

c. By the provision of a coarse graded slurry seal over an Asphalt or Coated Macadam surfacing. This is the least favoured option.

1.2.3 Runway Ends and Adjoining Aircraft Holding Areas

Service conditions can be severe in these areas where jet blast and fuel spillage are likely to be most damaging and aircraft carry out sharp turns. To avoid excessive tyre wear being caused to turning aircraft, it is usual to reduce the friction properties at runway ends compared with that provided on the main lengths. The material most frequently used on the runway ends and holding areas is pavement quality concrete. However, with certain aircraft the effect of jet blast on the pavement surface is not critical (e.g. piston-engined aircraft, turbo props or turbo jets where the jet exhaust is located at a safe height above the pavement). In these instances, bituminous surfacing provided with a fuel resistant surface treatment is likely to be a viable and economic alternative to pavement quality concrete.

1.2.4 Taxiways

Friction characteristics and rideability are not as critical as for the main length of a runway. Fuel and jet blast are not normally a critical consideration in these areas and therefore bituminous surfacings are generally a viable and economic construction option.

1.2.5 Parking and Servicing Aprons

Friction characteristics and rideability are not as critical as for the main length of a runway. In general, the need for resistance to fuel and oil spillages and to indentation by high tyre pressure tyres of standing aircraft and to damage from dropped tools, etc, will be such as to require the provision of a concrete surface. In certain circumstances, however, where the frequency of use is very low, bituminous surfacings provided with a fuel resistant surface treatment may be a viable and economic alternative to a concrete surface.

1.2.6 French Drains Adjacent to Airfield Pavements

Where French drains are provided adjacent or in proximity to airfield pavements, their surfaces must be stabilised in order to safeguard against the risk of FOD to aircraft. This can be achieved by the provision of a 100 mm thick Pervious Macadam topping to French drains.

1.3 Pavement Material Selection Principles

1.3.1 Unless the severity of service use of a pavement demands the use of concrete, bituminous surfacing material will normally be used.

1.3.2 The principal bituminous surfacing material used on MOD Airfields is Marshall Asphalt wearing and base course. Marshall Asphalt is a highly controlled and consistent material, provides a high stability and meets the specialist performance requirements for most airfield pavement applications. To aid proper control and make sure that the performance criteria will be met Marshall Asphalt is, wherever physically and economically possible, mixed on site. A separate DWS Functional Standard entitled "*Marshall Asphalt for Airfield Pavement Works*" gives details.

1.3.3 When either a new or existing airfield pavement is to be provided with a bituminous surfacing, apart from specialist surface treatments (e.g. Friction Course), Marshall Asphalt will be the preferred choice of material. However, there are circumstances in which, for economic and practical reasons, it will be preferable to use an alternative bituminous material. Such circumstances might include the following:

a. A small Scope of Work.

b. At Airfields where aircraft usage is low.

c. At Airfields not subject to regular trafficking by aircraft with high tyre pressures.

1.3.4 Further comment is given on the application of these factors at Clause 1.4. generally, they should be considered in combination rather than separately. Thus, if c. above applies but the volume of material is large enough to justify on-site mixing, then Marshall Asphalt would normally be the appropriate material to use.

1.3.5 In general, use of Hot Rolled Asphalt and/or Coated Macadam is applicable as a cheaper and more easily provided alternative for limited, non-critical portions of aircraft operating surfaces.

1.4 Application and Limitations of this Standard

1.4.1 Specifically, the use and specification for bituminous surfacing for aircraft pavement works must meet the functional requirements of the pavement concerned, and therefore careful evaluation is necessary before a decision can be made.

1.4.2 The specification requirements in this Standard provide an alternative to Marshall Asphalt for the situations outlined in sub-Clause 1.3.3. Specification requirements are provided for Hot Rolled Asphalt and Coated Macadam based on BS 594 and BS 4987 respectively but with additional requirements to meet the specialist performance requirements of airfield pavements. Specification requirements for Pervious Macadam 'topping to French drains' are also provided in this Standard.

1.4.3 This Standard is not a substitute or replacement for Marshall Asphalt wearing and base courses. However, as outlined in Clause 1.3, it provides alternative specification requirements for bituminous surfacing which should, in certain circumstances, be a more practical and economic solution than Marshall Asphalt. Specification requirements for Coated Macadam wearing courses are included as an option for limited applications (i.e.

for use by aircraft with low tyre pressures). However, Macadam wearing courses are rarely used for airfield pavements because of poor durability and stability.

1.4.4 Runway Friction Requirements

The friction characteristics on the main length of a runway must meet the criteria in NATO Stanag 3634, Airport Services Manual Part 2 (ICAO) and Annex 14 Volume 1 (ICAO). Previous experience of works on MOD runways has shown that plain Asphalt wearing courses rarely meet the friction criteria. Hence, prior to using the specifications in this Standard for runway resurfacing works, the Project/works Services Manager should first obtain advice from the Airfields and Bulk Fuels Group/DWS on grooving/texture requirements. Alternatives to texturing Hot Rolled Asphalt are given in sub-Clause 1.2.2 (a) and (c).

1.4.5 Limited Scope of Work

Marshall Asphalt involves special mix designs and laying methods and enhanced levels of control compared with conventional Hot Rolled Asphalt. Furthermore, to achieve the high quality control and consistency of production, it is normally a requirement to provide an on-site mixing plant. Hence, for small works, Marshall Asphalt is likely to be an unrealistic proposition. The specification requirements in this Standard should, in these circumstances, provide a viable alternative to Marshall Asphalt and give a significant saving in cost and mobilisation time.

1.4.6 Type of Aircraft and Frequency of Usage

These two design parameters have a considerable effect on the performance of bituminous surfacings. At airfields where aircraft usage and aircraft tyre pressures are low, the need for a high performance specification as provided by Marshall Asphalt may be unjustified. In these circumstances, the specification requirements in this Standard are likely to provide a viable alternative to Marshall Asphalt and give a significant saving in cost and in construction mobilisation time. However, for large scale works at critical or busy airfields, notwithstanding usage being mainly by aircraft with low tyre pressures, the consistency and enhanced performance provided by Marshall Asphalt is to be preferred over the specification requirements in this Standard. Laboratory tests carried out as part of the development of this Standard and experience gained from highway works have shown that high stability, as required for airfield pavement surfacings, can be achieved by designed Asphalt mixes based on BS 594. However, there is a lack of experience of the performance of these Asphalt mixes when subject to regular usage by aircraft with high tyre pressures. It is, therefore, recommended that the bituminous surfacings in this Standard are limited to locations where usage by aircraft with high tyre pressures is low as defined by Clauses Z.1 and Z.2 of Appendix Z to this Standard.

1.4.7 Climate

The specification requirements in this Standard are not intended for use in hot climates.

1.5 Specification Clauses for Hot Rolled Asphalt and Coated Macadam

Specification clauses are contained in Sections 2 to 9 and Appendices A to D of this Standard with Guidance Notes given on suitable temperatures and wind speeds for laying in Appendix E. Guidance Notes for the Project/Works Services Manager on Quality Systems are given in Appendix Y and for the preparation of job specifications in Appendix Z.

1.6 Advice from Airfields and Bulk Fuels Group/DWS

Clauses 1.2, 1.3 and 1.4 provide general advice on the application of this Standard. However, having regard to the various design parameters affecting the choice of construction and specification, including scope of work, aircraft type and frequency of usage, location of pavement on an airfield, design life, timescale constraints and existing pavement constructions, the guidance notes cannot be exhaustive. Further advice on a project/works specific basis can be obtained from the Airfields and Bulk Fuels Group/DWS.

2 General

2.1 References

All references to British Standards and other documents given in this Specification refer to the editions as listed in the References at the end of this document unless otherwise stated.

2.2 Overall Requirements

2.2.1 Hot Rolled Asphalt and Macadam shall be specified, mixed and laid to the requirements of BS 594: Parts 1 and 2, and BS 4987: Parts 1 and 2, and subject to the overriding clauses in this Specification.

2.2.2 The requirements of this Specification for any material are arranged in the following parts for the relevant material type:

	Hot Rolled Asphalt	Macadam	Pervious Macadam
General	Section 2	Section 2	Section 2
Materials	Section 3	Section 3	Section 3
Design and Composition	Section 4	Section 5	Section 6
Plant and Workmanship	Section 7	Section 7	Section 7
Trials	Section 8	Section 8	-
Tests	Section 9	Section 9	Section 9
Magnesium sulphate soundness test	Appendix A		
Stripping test	Appendix B		
Maximum density & air voids content	Appendix C		
Straightedge tests	Appendix D		
Suitable temperatures & wind speeds for laying	Appendix E		

2.3 Use of Hot Rolled Asphalt, Coated Macadam and Pervious Macadam

Hot Rolled Asphalt wearing course, Coated Macadam wearing course, Coated Macadam basecourse and Pervious Macadam drain topping shall be used in the locations indicated on the project drawings.

NOTE. Advice for the Project/Works Services Manager on the selection of the materials for specific job specifications is given in Clauses Z.1 to Z.3 of Appendix Z.

2.4 Quality Assurance for the Supply of Bituminous Materials

2.4.1 All operations in the procurement of component materials and batching of bituminous materials shall be carried out by a Contractor (or Supplier on his behalf) who has a

Quality Assurance accreditation to BS/EN/ISO 9000 (formerly BS 5750) for those operations.

NOTE. Advice for the Project/Works Services Manager on Quality Systems is given in Appendix Y.

2.4.2 Each production unit or depot involved in the work shall be registered under a Quality Assurance scheme to BS/EN/ISO 9000. The Quality Policy Manual/s for the supply of component materials and batching of bituminous materials, together with other relevant records and certificates, are to be submitted at Tender Stage.

NOTE. The Project/Works Services Manager should provide a questionnaire requesting the details of information that are required; advice is given in sub-Clauses Y.5.2 and Y.6.4 of Appendix Y.

2.4.3 The Contractor shall be responsible for having all testing for the supply of bituminous materials carried out in accordance with the requirements of Section 9 and provide the Project/Works Services Manager with a written copy of the results in accordance with Clause 9.1.

2.4.4 All documentation relevant to the work, including records of temperature control during mixing and test results, shall also be available at the plant or the depot for inspection. The documentation, including worksheets, shall be stored in an easily retrievable form for a minimum of 3 years.

3 Constituent Materials

3.1 Aggregates, General

3.1.1 The Contractor shall inform the Project/Works Services Manager of the source and aggregate properties for each aggregate. The type of coarse or fine aggregate to be used for a particular material shall be as shown in the following table:

Material	Aggregates				
	Coarse			Fine	
	Crushed Rock	Slag	Gravel	Sand	Crushed Rock
Hot Rolled Asphalt	Yes	No	No	Yes	Yes
Macadam	Yes	*	*	Yes	Yes
Pervious Macadam	Yes	Yes	No	No	No

Yes Suitable for use in material.
No Not for use in material.
* Not for use in material when used for wearing course, but suitable when used for other courses.

3.1.2 Initial approval of aggregates shall be obtained from the Project/Works Services Manager before mixing starts; approval shall be based on results supplied to the Project/Works Services Manager of those tests listed in Clause 9.2 and carried out by the Contractor.

3.1.3 The Contractor (or Supplier on his behalf) shall have Quality Assurance registration to BS/EN/ISO 9000 (formerly BS 5750) with an appropriate scope of registration for the production and supply of aggregates.

NOTE. Advice for the Project/Works Services Manager on Quality Systems is given in Appendix Y.

3.1.4 Aggregates shall be clean, hard and durable as defined in Clauses 3.2 and 3.3. Aggregates shall not contain deleterious materials in such a form or in sufficient quantity to adversely affect the strength at any age or the durability of the surfacing, including resistance to frost. Examples of such deleterious materials include:

- clay, particularly as an adherent coating;
- mica, shale and other laminated materials; and
- coal and other organic impurities.

3.1.5 Aggregate soundness of each source shall be assessed using the Magnesium Sulphate Soundness Test in accordance with Appendix A on all fractions.

OR

The aggregates do not have to be assessed using the Magnesium Sulphate Soundness Test in accordance with Appendix A provided there is local evidence that the aggregate is sound.

NOTE. Project/Works Services Manager to select option for specific job specification; advice given in Clause Z.4 of Appendix Z.

3.2 Coarse Aggregates

The properties of the coarse aggregate, determined in accordance with the methods described in the relevant reference, shall fall within the limits shown in the following table:

Test Property	Test Reference	Aggregate type	Material: Hot Rolled Asphalt	Material: Macadam	Material: Pervious Macadam
Minimum Magnesium Sulphate Soundness Value	Appendix A	Each source	82	82	-
		Each fraction	70	70	-
Maximum Flakiness Index (%)	BS 812: Part 105.1	Crushed Rock	30	30	30
		Gravel	n/a	30	n/a
Maximum Aggregate Crushing Value (%)	BS 812: Part 110	Crushed Rock	30	30	30
		Gravel	n/a	25	n/a
Maximum Absorption (%)	BS 812: Part 2	All except slag	2	2	2
		Slag	n/a	4	4
Stripping	Appendix B	All	Not greater than 6 particles from a 150 particle test sample shall indicate evidence of stripping		
Minimum Polished Stone Value	BS 812: Part 114	Runway wearing course	*	*	-
		Taxiway wearing course	*	*	-
Maximum Sulphur Content (per cent)	BS 1047	Slag	n/a	2	2
Stability	BS 1047	Slag	n/a	Requirements to Appendix (A)	
Minimum Bulk Density (kg/m³)	BS 812: Part 2	Slag	n/a	1120	1120

-   No limit.

n/a Not applicable.

*   Project/Works Services Manager to provide value for specific job specification; advice given in Clause Z.5 of Appendix Z.

3.3 Fine Aggregates

3.3.1 Fine aggregates shall be:

- natural bank, river, dune, or pit sand;
- crushed rock; or
- blends of sand and crushed rock

and shall be free from loosely bonded aggregations and other foreign matter. Sea-dredged sand shall not be permitted.

3.3.2 The properties of the fine aggregate, determined in accordance with the methods described in the relevant reference, shall fall within the limits shown in the following table:

Test: Property	Test: Reference	Aggregate type	Material: Hot Rolled Asphalt	Material: Macadam	Material: Pervious Macadam
Minimum Magnesium Sulphate Soundness Value	Appendix A	Each source	82	82	-
		Each fraction	70	70	-
Maximum Absorption (%)	BS 812: Part 2, Clause 5	All	2	2	2
Stripping	Appendix B	Parent rock if crushed rock fines	Not greater than 6 particles from a 150 particle test sample shall indicate evidence of stripping		

- No limit.

3.4 Filler

3.4.1 The Contractor (or Supplier on his behalf) shall have Quality Assurance registration to BS/EN/ISO 9000 (formerly BS 5750) with an appropriate scope of application for the production and supply of filler.

NOTE. Advice for the Project/Works Services Manager on Quality Systems is given in Appendix Y.

3.4.2 The type of filler to be used for a particular material shall be selected from the alternatives in the following table:

Material	Filler Type			
	Portland Cement (BS 12) *	Crushed Limestone	Crushed Rock	Hydrated lime (BS 890)
Hot Rolled Asphalt	Yes	Yes	Yes	No
Macadam	Yes	Yes	Yes	Yes
Pervious Macadam	Yes	No	No	Yes

Yes Suitable for use in material.
No Not for use in material.
* To be manufactured by a BSI Registered Firm.

3.4.3 Initial approval of filler shall be obtained from the Project/Works Services Manager before mixing starts; approval shall be based on results supplied to the Project/Works Services Manager of those tests listed in Clause 9.2 and carried out by the Contractor.

3.4.4 Filler shall be stored in dry conditions.

3.4.5 Filler gradings related to particular materials shall be in accordance with the following table:

BS 410 Sieves (μm)	Percentage by Mass Passing	
	Minimum	Maximum
300	100	-
75	*	100

- No limit.
* Project/Works Services Manager to provide value for specific job specification; advice given in Clause Z.6 of Appendix Z.

3.4.6 The bulk density of the filler, other than hydrated lime, in toluene determined according to BS 812: Part 2 shall be between 0.5 and 0.9 g/mℓ.

3.4.7 A copy of all filler delivery tickets shall be passed to the Project/Works Services Manager on a regular basis during production, for his retention.

3.5 Binder

3.5.1 The binder to be used for a particular material shall be in accordance with the following table:

Material	Binder Type Bitumen (BS 3690: Part 1) Penetration Grade			
	50	70	100	200
Hot Rolled Asphalt	Yes	Yes	Yes	No
Macadam — Wearing course	No	No	Yes	Yes
Macadam — Basecourse	Yes	No	Yes	No
Pervious Macadam	No	No	No	Yes

Yes Suitable for use in material/course.
No Not for use in material/course.

3.5.2 The particular grade of bitumen for a material shall be selected based on a number of factors in accordance with the requirements of Sections 4, 5 and 6 for Hot Rolled Asphalt, Macadam and Pervious Macadam, respectively.

3.5.3 The binder shall be penetration grade petroleum bitumen meeting the requirements of Table 1 of BS 3690: Part 1. The Contractor (or Supplier on his behalf) shall have Quality Assurance registration to BS/EN/ISO 9000 (formerly BS 5750) with an appropriate scope of application for the supply of bitumen.

NOTE. Advice for the Project/Works Services Manager on Quality Systems is given in Appendix Y.

3.5.4 Copies of certificates for the binder shall be passed to the Project/Works Services Manager for his retention.

3.6 Tack Coat

Tack coat shall be bitumen emulsion complying with either A1-40 or K1-40 of BS 434: Part 1.

3.7 Course Thickness

The thickness of each course of surfacing shall be as shown on the drawings. It shall be the thickness of the course at any point after compaction. Where the course thickness exceeds the layer thickness allowed in Clause 3.8, the course shall be laid and compacted in two or more layers.

NOTE. Increasing the thickness of a mat can considerably increase the time that it will take to cool to a temperature below which it cannot be effectively compacted. Therefore, in adverse weather (high winds and/or cold temperatures), the use of a greater course thickness will increase the possibility of having sufficient time to complete compaction.

3.8 Nominal Size of Aggregate Grading

The nominal size of aggregate grading for the particular material and nominal layer thickness to be used shall be in accordance with the following table:

Surfacing Material		Nominal Layer Thickness (mm)	Nominal Size of Aggregate Grading (mm)
Hot Rolled Asphalt		0 - 30‡ 30* - 45† 45★ - 50#	3 10 14
Macadam	Wearing course	40* - 50	14
	Basecourse	50† - 60 65 - 75	20 28
	Regulating and ramping	0 - 30 20 - 40 30 - 70 40 - 100	3 6 20 28
Pervious Macadam		100	28

* Thickness may be reduced to 20 mm when used for regulating and ramping.

★ Thickness may be reduced to 25 mm when used for regulating and ramping.

‡ For use in regulating and ramping only.

† Thickness may be increased to 50 mm when used for regulating and ramping.

\# Thickness may be increased to 60 mm when used for regulating and ramping.

4 Design and Composition of Hot Rolled Asphalt

4.1 General

4.1.1 Hot Rolled Asphalt wearing course shall be mixed and laid according to the requirements of BS 594: Parts 1 and 2, and subject to the overriding clauses in this Specification. The production of Hot Rolled Asphalt shall be carried out by a Contractor (or Supplier on his behalf) who has Quality Assurance registration to BS/EN/ISO 9000 (formerly BS 5750) with an appropriate scope of application for those operations.

NOTE. Advice for the Project/Works Services Manager on Quality Systems is given in Appendix Y.

4.1.2 The composition of Hot Rolled Asphalt wearing course mixtures shall be selected from the alternatives in Clauses 4.4 and 4.5, according to the specified thickness and the following table:

Location	Recipe Mixtures (Type B)	Design Mixtures		
		Type A	Type B	Type A or B
‡, *	†, *	†, *	†, *	†, *

‡ Project/Works Services Manager to state location(s) with drawing references as appropriate.

† Project/Works Services Manager to either tick the relevant box or state any specific requirements (e.g. bitumen grade or BS 594: Part 1 references for Type B Mixtures).

* Advice for Project/Works Services Manager given in Clauses Z.1, Z.7 and Z.8 of Appendix Z.

4.2 Design Mixtures

4.2.1 The composition of design mixtures shall comply with the requirements for either a Type A or a Type B mixture as directed by sub-Clause 4.1.2.

4.2.2 Design Hot Rolled Asphalt mixes shall be designed by the Contractor in his laboratory, using the size of aggregate grading appropriate to the layer thickness as specified in Clause 3.8, in accordance with BS 598: Part 107 to have a minimum stability of *.

* Project/Works Services Manager to provide value for specific job specification; advice given in Clause Z.8 of Appendix Z.

NOTE 1. Where 40 mm thick wearing course or thinner is specified, a binder of greater penetration than 50 pen bitumen is recommended provided the minimum stability requirement can be achieved. Extra workability will be gained and more time for compaction will be available.

NOTE 2. If it is necessary to increase the binder content to more than the optimum binder content to achieve the target binder content with an air voids content of less than 4.0 per cent, the stability value at the target binder content shall meet the minimum requirement above.

NOTE 3. The target binder content should be pitched such that, at the binder content of target + 0.4 per cent, the minimum stability value can be achieved and, at the target - 0.4 per cent binder content, the design air voids does not exceed 4 per cent.

4.2.3 At the target binder content for design mixes, the design air voids content determined in accordance with Appendix C shall be between 2.0 per cent and 4.0 per cent.

NOTE. A well compacted mat having a design air voids content of less than 4 per cent is expected to provide good durability.

4.2.4 Full details of the design mix shall be provided to the Project/Works Services Manager at least 1 week prior to laying the surfacing (or the laying trials). Once the Project/Works Services Manager has approved the mix design and proposed target binder content, the tolerance on the target binder content, on analysis, shall be ± 0.4 per cent.

4.2.5 Plant mixes shall be manufactured using the same aggregate, filler and binder as used in the design test.

4.3 Recipe Mixtures

4.3.1 Recipe mixtures shall be used at the locations permitted in sub-Clause 4.1.2.

4.3.2 Recipe mixtures shall be manufactured according to the relevant BS 594 aggregate grading and binder content, and subject to any additional Clauses in this Specification.

4.4 Composition of Hot Rolled Asphalt Mixture Type A

4.4.1 The Type A mixture is a design mixture in accordance with Clause 4.2 and which complies with the following table:

Component or Property	Requirement			
Filler	Portland cement or limestone; at least 60 per cent of the material passing the 75 micron sieve shall consist of added filler			
Aggregate grading	BS Sieve (mm)	Per Cent by mass passing	BS Sieve (μm)	Per Cent by mass passing
	14 10 6.3 2.36	100 70 - 100 60 - 78 45 - 57	600 212 75	20 - 36 10 - 21 3 - 6
Binder	70 pen bitumen or 100 pen bitumen †			
Target Binder Content	Determined in accordance with the criteria in Clause 4.2 but not less than 5.4 %			
Course thickness	40 mm to 45 mm (wider limits may be used for regulating and ramping, see Clause 3.8)			

† Use of 100 pen bitumen will improve workability at the expense of some stability.

4.4.2 The limits for binder content on analysis shall be ± 0.4 per cent.

4.5 Composition of Hot Rolled Asphalt Mixture Type B

4.5.1 The following BS 594 mixtures may be used subject within the restrictions described:

Course	Thickness (mm)	Bitumen grade	BS 594: Part 1 Reference
Wearing Course	30 - 40	50, 70 or 100 pen *	Table 6 Col 6/3 designation 30/10 †
	40 - 45	50, 70 or 100 pen *	Table 6 Col 6/4 designation 30/14 †
	45 - 50	50, 70 or 100 pen *	Table 6 Col 6/5 designation 40/14 †
	40 - 45	50, 70 or 100 pen *	Table 3 Col 3/4 designation 55/10 ‡
	45 - 50	50, 70 or 100 pen *	Table 3 Col 3/5 designation 55/14 ‡
Regulating and Ramping	0 - 30	50, 70 or 100 pen *	Table 6 Col 6/1 designation 0/3
	20 - 50	50, 70 or 100 pen *	Table 2 Col 2/1 designation 55/10
	20 - 50	50, 70 or 100 pen *	Table 3 Col 3/4 designation 55/10
	25 - 60	50, 70 or 100 pen *	Table 2 Col 2/2 designation 55/14
	25 - 60	50, 70 or 100 pen *	Table 3 Col 3/5 designation 55/14

‡ Design mixtures to Clause 4.2; **NOT TO BE LAID BY HAND.**

† Recipe mixtures to Clause 4.3; use only were permitted in accordance with sub-Clause 4.1.2.

* The use of bitumen with higher penetration will improve workability and widen the compaction window, at the expense of some stability.

NOTE 1. 100 pen bitumen is recommended in order to provide good workability. In many instances, it should be feasible to meet a 7.5 kN stability requirement using a 100 pen bitumen.

NOTE 2. Mixes 55/10 and 55/14 tend to segregate when hand laid, and mini-pavers should be used if small or intricate areas have to be laid. If hand laying cannot be avoided, mixes 30/14 or 40/14 should be used for that part of the work.

4.5.2 The following limits shall be applied to the design mixture composition on analysis:

- The binder content limits shall be ± 0.4 per cent.
- The aggregate grading limits for each sieve size shall be restricted from that given in BS 594: Part 1 as follows:

Difference between maximum and minimum per cent passing given in BS 594: Part 1	Restriction *
Less than 7	No change.
7 to 14	Increase the lower limit and reduce the upper limit by 1 per cent passing.
15 to 30	Increase the lower limit and reduce the upper limit by 2 per cent passing.
More than 30	Increase the lower limit and reduce the upper limit by 3 per cent passing.

* Where the upper limit is 100 per cent passing, no reduction should be made.

4.6 Mixing, Delivery and Compaction Temperatures

4.6.1 Hot Rolled Asphalts shall be mixed, delivered, laid and compacted within the material temperature limits given in the following table:

Course	Bitumen Grade	Temperature (°C)			
		Mixing (Max)	Delivery (Min)	Recommended Paver-out (Min)*	Compaction (Min)‡
Wearing Course	50 pen	185	140	130	100
or	70 pen	180	135	125	95
Regulating/Ramping	100 pen	170	130	120	85

* These values are required in order to achieve the maximum available compaction time. They are useful for monitoring purposes to ensure that adequate compaction time is available.

‡ This is the mid-layer temperature at which completion of compaction should have been achieved.

4.6.2 The Contractor shall check the temperature of the delivered load and the load in the hopper according to the method in BS 598: Part 109 at the following intervals:

- whilst discharging from the delivery lorry into the paver hopper;
- immediately before restarting the spreader following stoppage; and
- at any time the Project/Works Services Manager or his representative directs.

4.6.3 The rolling procedure shall have been completed before the surfacing temperature has fallen to the minimum compaction temperature.

5 Design and Composition of Macadam

5.1 General

5.1.1 Macadam basecourse and wearing course shall be mixed and laid according to the requirements of BS 4987: Parts 1 and 2, and subject to the overriding clauses in this Specification. The production of Macadam shall be carried out by a Contractor (or Supplier on his behalf) who has Quality Assurance registration to BS/EN/ISO 9000 (formerly BS 5750) with an appropriate scope of application for those operations.

NOTE. Advice for the Project/Works Services Manager on Quality Systems is given in Appendix Y.

5.1.2 Mixtures shall be manufactured according to the relevant BS 4987 aggregate grading and binder content, and subject to any additional clauses in this Specification.

5.2 Composition of Macadam Mixtures

5.2.1 Surfacing Macadams shall be selected from those of the following alternatives according to the specified thickness:

Course	Thickness (mm)	Bitumen grade	BS 4987: Part 1 Reference
Basecourse	50 - 60 65 - 75	* *	Clause 6.5, 20 mm size * Clause 6.4, 28 mm size *
Wearing Course	40† - 50	100 pen (or 200 pen ‡)	Clause 7.3, 14 mm close graded
Regulating and Ramping	0 - 30 20 - 40 30 - 70 40 - 100	100 pen (or 200 pen ‡) 100 pen 50 pen or 100 pen 50 pen or 100 pen	Clause 7.7, 3 mm fine graded Clause 7.5, 6 mm dense wearing course Clause 6.5, 20 mm dense basecourse Clause 6.4, 28 mm dense basecourse

* Project/Works Services Manager to provide bitumen grade and material type (DBM, DBM 50 or HDM) for specific job specification; advice given in Clause Z.9 of Appendix Z.

‡ Project/Works Services Manager to delete unless required for specific job specification; advice given in Clause Z.9 of Appendix Z.

† Thickness may be reduced to 20 mm when used in ramps.

5.2.2 The following limits shall be applied to the design mixture composition on analysis:

- The binder content limits shall be $\pm$ 0.4 per cent.
- The aggregate grading limits for each sieve size shall be restricted from that given in BS 4987: Part 1 as follows:

Difference between maximum and minimum per cent passing given in BS 4987: Part 1	Restriction *
Less than 7	No change.
7 to 14	Increase the lower limit and reduce the upper limit by 1 per cent passing.
15 to 30	Increase the lower limit and reduce the upper limit by 2 per cent passing.
More than 30	Increase the lower limit and reduce the upper limit by 3 per cent passing.

* Where the upper limit is 100 per cent passing, no reduction should be made.

5.3 Mixing, Delivery and Compaction Temperatures

5.3.1 Surfacing Macadams shall be mixed, delivered, laid and compacted within the material temperature limits given in the following table:

Course	Bitumen Grade	Temperature (°C)			
		Mixing (Max)	Delivery (Min)	Recommended Paver-out (Min)*	Compaction (Min)‡
Basecourse	100 pen	160	120	110	85
Basecourse DBM 50	50 pen	165	130	120	95
Basecourse HDM	50 pen	170	150	140	115
Wearing course	200 pen	150	115	105	75
Wearing course	100 pen	160	120	110	85
Regulating/Ramping	100 pen	160	120	110	85

* These values are required in order to achieve the maximum available compaction time. They are useful for monitoring purposes to ensure that adequate compaction time is available.

‡ This is the mid-layer temperature at which completion of compaction should have been achieved.

5.3.2 The Contractor shall check the temperature of the delivered load and the load in the hopper according to the method in BS 598: Part 109 at the following intervals:

- whilst discharging from the delivery lorry into the paver hopper;
- immediately before restarting the spreader following stoppage; and
- at any time the Project/Works Services Manager or his representative directs.

5.3.3 The rolling procedure shall have been completed before the surfacing temperature has fallen to the minimum compaction temperature.

6 Design and Composition of Pervious Macadam

6.1 General

The production of Pervious Macadam shall be carried out by a Contractor (or Supplier on his behalf) who has Quality Assurance registration to BS/EN/ISO 9000 (formerly BS 5750) with an appropriate scope of application for those operations.

NOTE. Advice for the Project/Works Services Manager on Quality Systems is given in Appendix Y.

6.2 Design Mixtures

6.2.1 The target binder content shall be calculated in accordance with the Binder Drainage Test in Clause 939 of the *Specification for Highway Works*, Volume 1 of the *Manual of Contract Documents for Highway Works* (Her Majesty's Stationery Office) except that the binder contents used shall be 2.5, 3.0, 3.5, 4.0 and 4.5 per cent by mass respectively.

6.2.2 The Binder Drainage Test shall be carried out at:

- 125° ± 3°C when using 200 pen bitumen;
- 150°C ± 3°C when using natural rubber latex or powder; and
- in accordance with the manufacturer's recommendations for other modifiers.

For other modifiers, the base bitumen used shall be 200 pen bitumen.

6.2.3 The Binder Drainage Test shall be carried out at the mid-point of the grading limits using the same coarse aggregates and filler option as proposed for use in the Works.

NOTE 1. Aggregates vary in their shape and surface characteristics such as binder absorption. The Binder Drainage Test determines the maximum binder content that can be safely used without excessive binder drainage.

NOTE 2. Non-proprietary modifiers such as natural rubber latex or powder may be used. The Binder Drainage Test in Clause 939 gives details of the test using these modifiers.

NOTE 3. Proprietary modifiers or modified binders may be used subject to a satisfactory result in the Binder Drainage Test. The procedure for incorporating the modifier shall be according to the manufacturer's instructions. Full details should be supplied to and advice sought from the Defence Works Services prior to approval to use a proprietary system being given.

6.2.4 The target binder content shall be the lesser of the calculated target binder and 4.5 per cent. The mixture shall not be used where:

- the target binder content is less than 3.0 per cent using unmodified 200 pen bitumen; or
- the target binder content is less than 4.0 per cent using a modified binder.

The manufacturing tolerance on the target binder content shall be ± 0.3 per cent.

6.2.5 The Contractor shall provide full details of the Binder Drainage Test to the Project/Works Services Manager.

6.2.6 Plant mixes shall be manufactured using the same aggregate, filler and binder as used in the binder drainage test. The mixing temperature shall not exceed that specified in the Binder Drainage Test.

6.3 Recipe Mixtures

Recipe mixtures shall use 200 pen unmodified bitumen with a target binder content of 3.4 per cent when the coarse aggregate is crushed rock and 4.3 per cent when it is slag. The manufacturing tolerance on the target binder content shall be ± 0.3 per cent.

6.4 Composition of Pervious Macadam

6.4.1 The aggregate grading shall comply within the limits of the following table. The Contractor shall target at the mid-point of the grading limits.

BS Sieve (mm)	Per cent passing
37.5	100
28	80 - 90
20	10 - 30
14	4 - 6

6.4.2 Fine aggregate shall not be added.

6.4.3 The mixture shall contain by mass of the total mixture either:

- 2 per cent of hydrated lime;
- 4 per cent of Ordinary Portland Cement, by mass of the total aggregates; or
- Up to 1.5 per cent of Stearine Amine or other approved wetting agent by mass of the total binder content so long as it can be demonstrated to the approval of the Project/Works Services manager that no drainage of the binder from the aggregate takes place during transport to the point of placing.

NOTE 1. Project/Works Services Manager to delete final option when not required for specific job specifications; advice given in Clause Z.10 of Appendix Z.

NOTE 2. The Contractor shall state which filler option he chooses.

6.4.4 Depending on the application, the binder selected shall be either:

- 200 pen bitumen; or
- 200 pen bitumen with an approved modifier.

Pervious Macadam using unmodified 200 pen bitumen shall only be used for small works and for remedial works to French drain toppings with the approval of the Project/Works Services Manager.

NOTE. Pervious Macadam using 200 pen bitumen and a modifier will have a higher binder content than unmodified Pervious Macadam, and is expected to have improved durability; it should be used for new works.

6.5 Mixing, Delivery and Compaction Temperatures

6.5.1 Pervious Macadams shall be mixed, delivered, laid and compacted within the material temperature limits given in the following table:

Bitumen Grade	Temperature (°C)			
	Mixing (Max)	Delivery (Min)	Recommended Paver-out (Min)*	Compaction (Min)‡
200 pen 200 pen plus natural rubber	125	95	85	65
200 pen plus other modifier	†	†	†	†

† In accordance with the Supplier's recommendation.

* These values are required in order to achieve the maximum available compaction time. They are useful for monitoring purposes to ensure that adequate compaction time is available.

‡ This is the mid-layer temperature at which completion of compaction should have been achieved.

6.5.2 The Contractor shall check the temperature of the delivered load and the load in the hopper according to the method in BS 598: Part 109 at the following intervals:

- whilst discharging from the delivery lorry; and
- at any time the Project/Works Services Manager or his representative directs.

6.5.3 The rolling procedure shall have been completed before the surfacing temperature has fallen to the minimum compaction temperature.

7 Plant and Workmanship

7.1 General

7.1.1 Mixtures shall be transported, laid and compacted according to the BS 594 for Hot Rolled Asphalt and BS 4987 for Coated Macadam, and subject to any additional clauses in this Specification.

7.1.2 The standard of workmanship and finish of all surfacing included in this Contract shall be equal in all respects to that of the "Approved" areas established in the trials in accordance with Section 8.

7.2 Weather Conditions for Laying

7.2.1 Laying of bituminous surfacing shall not proceed unless:

- the surface to be covered is unfrozen and free from ice, snow and deicing agents;
- the temperature of the surface to be covered is 0°C or more; and
- the air temperature is either:
  - above 1°C or
  - between -1°C and 1°C and rising.

7.2.2 Laying of bituminous surfacing shall not proceed during precipitation unless:

- both the surface to be covered and the air temperature are above 0°C;
- there is no free water on the surface; and
- the degree of moisture present on the surface is not detrimental to the finished product.

NOTE. Guidance is given on suitable temperatures and wind speeds for laying in Appendix E.

7.3 Reduction in Surface Level of Bituminous Surfaces

Where the surface level of an existing bituminous surface is being reduced, thicknesses shall be removed with an approved planing machine. The machine shall be provided with control devices which enable the rapid adjustment of blades to fine depth-of-cut settings while the machine is operating.

7.4 Reduction in Surface Level of Concrete Surfaces

7.4.1 Where the surface level of an existing concrete surface is being reduced, thicknesses from fine fractions up to the limits as specified hereafter shall be removed by scabbling or planing. Scabbling or planing shall be carried out by machines operated in a manner that results in a minimum of over-cutting and uniform exposure of the aggregate, without shattering or otherwise damaging the concrete slabs.

7.4.2 Unless directed by the Project/Works Services Manager, such reductions in level shall be bounded by joints in the concrete pavement. Where they are not and where the scabbling or planing is not tapered to zero depth, the area to be reduced in level shall first be defined by saw cuts, 5 mm deeper than the depth of the scabbling or planing.

7.4.3 The surface treated as above shall be thoroughly cleaned with hand brooms and all loose debris shall be collected and removed.

7.4.4 Before the area is surfaced, a tack coat appropriate to the surfacing material shall be applied as specified in Clause 7.9 and the exposed vertical edges shall be painted with hot bitumen. The new surfacing shall be made to firmly adhere to the vertical edges.

7.4.5 Where the thickness to be removed is greater than 100 mm or one third the depth of the slab, whichever is less, the whole of the concrete layer shall be removed for its full depth. Where the concrete is in bays less than 4.5 m square, the boundary of the area to be removed shall be defined by bay joints. Where the bays are larger, part bays may be removed such that the remaining portion shall not be less than 2.25 m wide by the full length of the bay or 2.25 m long by the full width of the bay.

7.4.6 Care shall be taken to ensure that all concrete remaining in the vicinity of cutting-out remains sound and without fracture and that disruption to sub-bases and underlying layers is minimised.

7.5 Preparation of Existing Bituminous Surfaces

7.5.1 Before the tack coat is applied, all vegetable growth and loose aggregate or other particles shall be removed from all cracks in the existing bituminous surfacing on which new surfacing is to be laid. Loose laitance and other spalling or debonded slurry seal shall be removed from the surface. The surfaces shall be swept until standing water, mud, grit and all other extraneous matter has been removed. Immediately ahead of tack coating, all dust shall be removed by vacuum extraction cleaning, with or without high pressure water at the discretion of the Project/Works Services Manager.

NOTE 1. Cleaning machines applying high pressure water and vacuum extraction are very effective in removing debris and cleaning the surface. Water pressures about 50 Bar are adequate for general cleaning, but higher pressures may be used subject to the Project/Works Services Manager's discretion. At very high pressures, damage to the surfacing can result.

NOTE 2. Vacuum extraction after high pressure cleaning generally results in a sufficiently dry surface for application of a tack coat.

7.5.2 Existing overbanding of the surfacing shall be removed when required by the Project/Works Services Manager.

NOTE. When resurfacing comprises only a wearing course, it will normally be a requirement to at least remove thick depositions of overbanding.

7.6 Preparation of Existing Concrete Surfaces

7.6.1 Before the tack coat is applied, all vegetable growth and loose debris including concrete fractions shall be removed from all cracks, joints and joint edges in the existing concrete surfacing on which the new bituminous surfacing is to be laid. The surface shall be thoroughly cleaned and all loose debris shall be collected and removed.

NOTE 1. Cleaning machines applying high pressure water and vacuum extraction are very effective in removing debris and cleaning the surface. Water pressures about 50 Bar are adequate for general cleaning, but higher pressures may be used subject to the Project/Works Services Manager's discretion. At very high pressures, damage to the surfacing can result.

NOTE 2. Vacuum extraction after high pressure cleaning generally results in a sufficiently dry surface for application of a tack coat.

7.6.2 All joints in the concrete paving shall be examined and a joint preparation programme shall be agreed between the Contractor and the Project/Works Services Manager.

7.6.3 When directed, in joints 20 mm wide or greater where the joint sealing compound has lost its original properties and can be removed by hand methods, it shall be replaced with lightly compacted 3 mm size fine graded wearing course in accordance with Clause 7.7 of BS 4987: Part 1, laid to a convex finish slightly proud of the general surface level.

7.6.4 In all other cases, the joints shall be prepared by cutting-off any compound which has extruded above the general level of the pavement with heated tools so as to be flush with the concrete surface. Any unfilled joints less than 20 mm wide may be ignored.

7.6.5 Loose laitance and other spalling or debonded slurry seal shall be removed from the surfaces which shall then be swept until standing water, mud, grit and all other extraneous matter has been removed. Immediately ahead of tack-coating, all dust shall be removed by vacuum extraction cleaning, with or without high pressure water at the discretion of the Project/Works Services Manager.

7.7 Filling Ravelled Lane Joints, Ravelled Cracks and Potholes in Existing Bituminous Surfacing

7.7.1 Ravelled joints, ravelled cracks and potholes shall be made good before the new surfacing is laid.

7.7.2 Trenches shall be formed by carefully cutting out the existing bituminous surfacing on either side of the joints or cracks to the full depth of the wearing course and, if directed, to the underside of the basecourse or to the top of the underlying concrete or pavement base. The new material shall be bonded into the old surfacing.

7.7.3 The cross section of the trench shall be a minimum of 200 mm wide. The side walls of the trench shall be clean vertical cuts and shall be stepped-back a minimum of 50 mm on each side at a convenient plane of separation between any two courses of the existing surfacing. When the existing pavement level is not to be raised, the edges of the trench or patch shall be defined by means of saw cuts extending to the full depth of the wearing course.

7.7.4 All loose and crumbling fractions shall be removed from the bottom and sides of the trench. The bottom and sides shall be completely painted with tack coat.

7.7.5 The defective surfacing shall be replaced with the specified wearing course material. It shall be placed in the trenches in lifts of about 50 mm each which shall be compacted separately with approved mechanical or hand tampers as specified in Clauses 7.15 and/or 7.16.

7.7.6 At the time of compaction, the mixture shall be at the specified temperature. The final layer shall be laid so as not to leave a concave finish below the general surface after thorough compaction by rolling.

7.7.7 After removing all loose material from the potholes, the Contractor shall paint the bottom and sides of the depressions with tack coat and backfill with wearing course material as specified above, placed, compacted and finished in a similar manner.

NOTE. Further guidance on making good is given in Functional Standard 06, "*Guide to Airfield Pavement Maintenance*" (Defence Works Services 1994).

7.8 Transporting Plant Mixtures

7.8.1 The plant mixtures shall be transported without delay to the laying sites from the mixing plant or from hot storage bins taking care to prevent segregation. The vehicles shall be double sheeted during transit and while waiting to prevent loss of heat, contamination and wetting. All vehicles shall be mechanically sound and shall be suitable for the spreading equipment in use and shall have insulated bodies.

7.8.2 The use of diesel oil or water on the surfaces of the transporting vehicles to facilitate discharge must be strictly regulated to the absolute minimum and, if the Project/Works Services Manager considers that contamination of the mixtures is occurring, the vehicle shall be thoroughly cleaned out to his satisfaction before being used again. The use of dust, sand or other fine particles is prohibited.

7.8.3 The temperature of the load in every transporting vehicle shall be checked in accordance with BS 598: Part 109 whilst discharging from the delivery lorry into the spreader. If the temperature of any batch is below the minimum delivery temperature specified in Clause 4.6 for Hot Rolled Asphalt, Clause 5.3 for Macadam or Clause 6.5 for Pervious Macadam, the load shall be rejected and shall be removed from site immediately. If the delivery temperature of Pervious Macadam is greater than 100°C when 200 pen bitumen has been used, or greater than 120°C if modified 200 pen bitumen has been used, the load shall be rejected and shall be removed from site immediately.

7.8.4 Each delivery of mixed material to the contract works, whether batched on-site or off-site, shall be accompanied by a delivery ticket giving the following details:

- Delivery ticket number;
- Vehicle registration number;
- Material type and mix classification;
- Nominal penetration grade of bitumen used; and
- Quantity.

For material batched off-site, the delivery ticket shall also give the following details:

- Customer name and Delivery site number;
- Source of supply;
- Date and time loaded; and
- Date and time delivered.

7.8.5 Copies of the delivery tickets shall be passed to the Project/Works Services Manager for his retention.

7.9 Tack Coat Application

7.9.1 The type of tack coat to be used for the particular material being laid is as specified in Clause 3.6.

7.9.2 Bitumen emulsion tack coat shall be applied not more than 48 hours in advance of surfacing:

- at a rate of between 0.35 ℓ/m^2 and 0.55 ℓ/m^2 to all the existing surface prior to laying of the new surfacing; and
- at a rate of between 0.25 ℓ/m^2 and 0.35 ℓ/m^2 between layers when the new surfacing includes 2 or more layers if:
    - o the binder content of the underlying layer is 5.5 per cent or less; or
    - o the underlying layer has been left exposed for more than 7 days.

Bitumen emulsion tack coat may be applied to damp surfaces but ponded or standing water shall be removed as specified in Clauses 7.5 and 7.6.

7.9.3 Tack coat shall be applied uniformly, free of streaks and blobs in accordance with BS 434: Part 2 by mobile mechanical tank-spraying units complying with BS 3136: Part 2. The tack coat shall be allowed to 'break' completely before laying proceeds. Where the size or shape of an area to be sprayed precludes mobile operation, pressure spraying equipment or hand-spraying complying in accordance with BS 434: Part 2 will be permitted with the approval of the Project/Works Services Manager.

NOTE. The use of paving machines that incorporate equipment to apply the tack coat immediately before the mix is laid will not allow the opportunity to ensure that the tack coat has 'broken'. If the Contractor wishes to use such equipment, he shall seek prior written approval from the Project/Works Services Manager. Advice for Project/Works Services Manager is given in Clause Z.11 of Appendix Z.

7.9.4 Airfield lighting units, gratings, covers and similar fittings shall be adequately masked with an approved protection during application. Care shall be taken to prevent the spraying of porous surfacing of the french drains and, if the Project/Works Services Manager considers it to be necessary, these too shall be protected.

7.9.5 After application, no traffic of any kind shall be allowed to run over the tack coat until surfacing starts and arrangements shall be made to cordon off the sprayed areas until it does. When surfacing starts, only the minimum amount of traffic essential to the surfacing operations shall be permitted.

7.10 Permitted Tolerance of Course Thickness

7.10.1 The total compacted thickness of any course of the surfacing material at any point shall not be less than the specified course thickness or exceed this thickness by more than the limits of the following table:

Surfacing Material	Maximum Permitted Tolerance on the Specified Course Thickness (mm)	
	Basecourse	Wearing Course
Hot Rolled Asphalt	n/a	+ 10* / - 0
Macadam	+ 25 / - 0	+ 10* / - 0
Pervious Macadam	n/a	+ 25 / - 0

* Except in ramps (see Clause 7.22).
n/a Not applicable.

7.10.2 Core samples shall be taken to determine course thickness:

- at minimum intervals of every 1000 m² laid or from every 2 hours work, whichever is the more frequent; and
- at locations agreed with the Project/Works Services Manager.

7.11 Regulation of Existing Surfaces

7.11.1 Where the irregularities in the pavements to be surfaced are such that the permitted thickness tolerances for that course will be exceeded, the existing surfaces shall be regulated as a separate item in advance of general resurfacing.

7.11.2 Regulation for depths less than 40 mm shall be carried out with a relevant mixture from Clause 4.4, 4.5 and/or Clause 5.2 subject to the following requirements:

- Regulation for depths 0 to 20 mm where it is to be immediately beneath a design mixture Hot Rolled Asphalt wearing course:
  - Hot Rolled Asphalt complying with Table 6, Column 6/1, designation 0/3 shall be used; and
  - The bitumen penetration grade shall be 70 unless the wearing course is a 10 kN design stability, when the penetration grade shall be 50.
- The positioning of regulating courses shall be planned to minimise the use of fine Hot Rolled Asphalt (Table 6, Column 6/1 of BS 594: Part 1).
- Regulation for depths 20 mm to 30 mm where it is to be immediately beneath a design mixture Hot Rolled Asphalt wearing course, Hot Rolled Asphalt complying with the same design stability requirement as the wearing course (i.e. as required by sub-Clause 4.2.2) shall be used.
- Regulation for depths 30 mm to 40 mm where it is to be immediately beneath a design mixture Hot Rolled Asphalt wearing course:
  - Hot Rolled Asphalt complying with the same design stability requirement as the wearing course (i.e. as required by sub-Clause 4.2.2); or
  - 20 mm DBM 50 or HDM Macadam basecourse complying with Clause 6.5 of BS 4987: Part 1

  shall be used.

7.11.3 Regulation for depths in excess of 40 mm shall be carried out with a relevant mixture from Clause 4.4, 4.5 and/or Clause 5.2 except that, where it is to be immediately beneath a design mixture Hot Rolled Asphalt wearing course, it shall be either:

- Hot Rolled Asphalt complying with the same design stability requirement as the wearing course (i.e. as required by sub-Clause 4.2.2);
- 20 mm DBM 50 or HDM Macadam basecourse complying with Clause 6.5 of BS 4987: Part 1; or
- 28 mm DBM 50 or HDM Macadam basecourse complying with Clause 6.4 of BS 4987: Part 1.

7.11.4 Regulation of the existing surface shall continue subject to the tolerances specified in Clause 7.10 until the regulated surface achieves an accuracy of:

- 10 mm in 3 m when a wearing course only is being laid; or
- 25 mm in 3 m when a basecourse is included in the new surfacing,

when the regulated surface is tested with a straightedge in the manner as detailed in Appendix D.

7.12 Laying Requirements

7.12.1 A competent supervisor shall be in charge of all laying and finishing operations.

7.12.2 The mixtures shall be spread to surcharged depths necessary to give the specified course thickness, and to comply with the finished levels and profiles shown on the drawings, after compaction.

7.12.3 Wearing course shall be laid on basecourses as soon as practicable in order to provide adhesion with the basecourse. Basecourses which have been left exposed for more than seven days shall be tack coated in accordance with Clause 7.9 before the laying of wearing course.

7.12.4 The use of hand-rakes shall be prohibited when the mixtures are laid by spreading and finishing machines in accordance with Clause 7.14 except at joint edges and around manholes and pits, where their use shall be restricted to an absolute minimum.

7.12.5 After the spreading units have passed, hand-casting of fines behind the spreader as a means of making-up irregularities or disguising blemishes left by the spreader shall not be permitted.

7.12.6 At all times, the courses shall be kept free from all extraneous matter.

7.13 Sampling of Mixed Materials

7.13.1 Bulk samples of the mixed material sent to the site, as required in sub-Clause 9.5.1, shall be taken by the Contractor and divided. One sample shall be analysed for grading and binder content, and the other retained for reference in the case of a dispute.

7.13.2 The samples can be taken at the batching plant or at the site, with the alternative selected to be notified at Tender Stage. If the sample is to be taken at the site, it shall be taken from the paver screws after approximately half of the lorry load has been discharged through the paver.

7.13.3 Samples shall be labelled and details shall include material type, date of delivery, vehicle registration number, course location and time of laying and other relevant information deemed necessary by the Project/Works Services Manager.

7.14 Spreading by Machine

7.14.1 Except where the conditions of Clause 7.16 apply, the mixture shall be spread, levelled and tamped by approved self-propelled spreading and finishing machines which are capable of continuously laying to the required widths, profile, camber or crossfall without causing segregation, dragging, burning or other surface defects or irregularities. They shall also be capable of operating at a speed consistent with the type and thickness of the Asphalt or Macadam being laid.

NOTE. The method of control should be adequate to achieve the tolerances required and should not be limited by the length of the paving equipment if that is not sufficient. It may be necessary to use a wire guidance system or averaging beam to achieve the required accuracy in certain critical situations, such as at or in the vicinity of wandering crowns or for laying of regulating courses.

7.14.2 Any extension beyond the basic width of the machine shall be strictly in accordance with the manufacturer's recommendations and shall give a level uniform surface over the full width of the lane to the satisfaction of the Project/Works Services Manager.

7.14.3 Each spreader shall be maintained in good mechanical condition and shall be correctly adjusted for operation at the speed consistent with the character and rate of delivery of the mixture and the thickness and agreed rolling procedures for the course, to produce a surface of uniform density and texture free from segregation, dragging, irregularities, or other unacceptable surface blemishes. If dragging or other faults should occur, laying shall cease until the mechanism and operation of the units have been checked and the defects have been rectified or modifications made.

7.14.4 As soon as possible after arrival at the laying site, the mixtures shall be discharged continuously to the spreader and shall be laid in accordance with the requirements of

Clause 7.12 without delay. When discharging into the spreader, the lorry shall approach and gentle contact shall be made only between rollers on the spreader and the rear wheels of the lorry to avoid causing the paver screed to indent the mat.

7.14.5 Intermittent stopping of the spreader shall be avoided and the rate of delivery to the spreader shall be so regulated to enable the spreader to be operated continuously.

7.15 Compaction

7.15.1 The surfacing shall be uniformly compacted in the manner approved during the laying of the trial area described in Clause 8.3, using the type of equipment and loads applied as agreed with the Project/Works Services Manager.

7.15.2 Rollers shall be in good condition and fitted with smooth rapid acting reverse controls. They shall be equipped with roll scrapers, absorbent mats and tanks connected to spray pipes on both front and rear rolls to ensure a uniform minimal application of water or parting fluid. The rollers shall be operated by skilled and experienced drivers. The weight to which each roller shall be ballasted shall be agreed with the Project/Works Services Manager during the laying of the trial(s).

7.15.3 Rolling shall proceed in the direction of laying with the rear wheel lapping the edge of any previously laid surfacing and shall progress gradually to the opposite edge of the lane. The lapping of the rolling shall be such that, on completion, all roller marks are obliterated. During rolling, the roller wheels shall be kept moist with only enough water to avoid picking up material. A water bowser shall be provided alongside each spreading unit to ensure that rolling continues with the minimum interruption.

7.15.4 Rollers shall move at a slow but uniform speed which should not exceed 5 km/h and any pronounced steering change in direction of the roller shall be made on stable material. The line of rolling shall not be suddenly changed or the direction of rolling suddenly reversed, thereby displacing the mix. Rollers shall not be left standing on the new surfacing within 24 hours of laying.

7.15.5 The roller sequence and types shall be such as to provide the required standard of compaction and finish.

7.16 Spreading and Compacting by Hand

7.16.1 Spreading by hand will be permitted for:

- filling potholes;
- the regulation of existing surfaces as specified in Clause 7.11;
- feathering;
- shaping drainage channels if impracticable by machine;
- the replacement of old or defective surfacing when the areas are small;
- areas of irregular shape; and
- laying Pervious Macadam to French drains.

With the approval of the Project/Works Services Manager, spreading by hand will also be permitted in areas where manholes or pits are concentrated and in areas which are inaccessible to the spreading and finishing machines specified in Clause 7.14.

7.16.2 The mixture shall be unloaded with care to avoid segregation onto an existing hard, clean surface adjacent to the area on which it is to be placed or, when this is not available, onto an approved metal sheet alongside the area. The mixture shall be spread portion by portion without break with hot shovels to a uniform thickness which, after compaction, shall not exceed the maximum thickness specified for the mixture. The material shall then be finished with hot hand-rakes by skilled rakers to the level required to give the correct shape and profile after compaction.

7.16.3 The exposed edges of manhole frames, grating frames, lighting units and any fixtures in the pavement or the concrete surrounds against which the new surfacing abuts shall be scraped and thoroughly cleaned to the satisfaction of the Project/Works Services Manager. An approved sealing system shall be applied around the fixture/surround in accordance with the manufacturer's instructions, or other treatment as recommended in BS 594 or BS 4987, within 2 hours prior to laying the bituminous surfacing. The surfacing shall then be packed tightly around the fixture and firmly tamped into position.

7.16.4 On completion of compaction, the finished surface of the wearing course and other materials as used shall be level with the fixture to the accuracy specified in Clause 7.25. Shaping and dishing to the sides of catchpits shall be carefully carried out to the profiles shown on the drawings. In places inaccessible to the rollers specified in Clause 7.15, compaction shall be achieved by suitable vibrating rollers or by tamping.

7.17 Topping to French Drains

7.17.1 Spreading of Pervious Macadam may proceed provided the weather conditions comply with Clause 7.2. The covers on the delivery lorry shall only be rolled back sufficiently to discharge sufficient material for immediate use. At the discretion of the Project/Works Services Manager, Pervious Macadam may be laid by a small width screed paver where support on either side of the French drain is deemed sufficient. Alternatively, Pervious Macadam shall be spread with hot shovels carefully to avoid contamination from adjacent vegetable growth or other foreign material. It shall be spread portion by portion in a single uniform layer along the length of the drain to the required width, and levelled off with hot hand rakes by skilled operators to give the correct shape and profile after compaction.

7.17.2 The Pervious Macadam shall be compacted by a smooth wheeled tandem roller of mass not exceeding 3 tonnes having a roller width not exceeding that of the French Drain. No vibration shall be permitted. Pervious Macadam shall be laid within 3 hours of manufacture.

7.18 Longitudinal Lane Joints, General

Longitudinal joints in surfacing materials shall be constructed in such a position that they are at least 600 mm horizontally away from any longitudinal joints in the underlying material. The longitudinal lane joints shall be vertical in straight lines which are continuous for the full length of the pavement, or in smooth curves around bends.

7.19 Longitudinal Lane Joints in Materials other than Pervious Macadam

7.19.1 The exposed vertical edges of the longitudinal lane joints in the surfacing materials shall be carefully cut back and trimmed to firm material in the compacted lane, or for a minimum of one half and a times the layer thickness, whichever is the greater, and all loose material arising from this operation shall be removed from the pavement before the cut edge is painted.

7.19.2 Edge rolling shall only be used as an alternative to cutting back if it can be demonstrated during the trials to the Project/Works Services Manager that satisfactory standards of compaction, surface/joint finish and adhesion can be achieved.

NOTE. Cores should be taken for test at these joints to demonstrate good adhesion and bulk density within the specified limits.

7.19.3 Cutting back and trimming will not be required when two or more spreading units operate in echelon in close proximity, permitting adjacent lanes to be continuously compacted before the material around the joint between the lanes falls below the compaction temperature specified in Clause 4.6 for Hot Rolled Asphalt or Clause 5.3 for Macadam.

7.19.4 After cutting back and trimming, the exposed vertical edges of the longitudinal lane joints shall be thoroughly cleaned of all adherent material and shall then be painted with a uniform thickness of hot bitumen just ahead of the spreading unit laying the adjacent lane. Painting shall completely and uniformly cover the exposed edge for its full depth. Excess material to the top and base of the joints, streakiness and blobs shall be avoided.

7.19.5 On completion, the joints shall present the same texture as the remainder of the surface and the accuracy of the surface across the joints shall meet the criteria specified in Clause 7.25.

7.19.6 When laying during cool windy weather, a joint heater may be used subject to the discretion of the Project/Works Services Manager.

7.20 Transverse Joints

7.20.1 Transverse joints are required at the end of a day's work and following any interruption in laying which prevents continuity of rolling at, or above, the specified minimum temperature. Transverse joints shall be formed at right angles to the longitudinal joints and shall be vertical.

7.20.2 The exposed vertical edges of the transverse joints of all layers shall be cut back for at least 300 mm and trimmed. Arisings from this operation shall be removed from the pavement and the underlying surface cleaned. The exposed joint edges shall then be cleaned and painted with hot bitumen as specified in Clause 7.19 immediately before the laying of the lane continues.

7.20.3 On completion, the joints shall present the same texture as the remainder of the surface and the accuracy of the surface across the joints shall meet the criteria specified in Clause 7.25.

7.21 Joints between New Surfacing and Existing Pavements

7.21.1 Existing bituminous surfacing against which new surfacing is to be laid shall be cut back as necessary to a line removing all loose or weathered material and shall be finished with a vertical edge. Immediately prior to the laying of new material, either:

- a thin uniform coating of 50 pen or 70 pen hot bitumen; or
- an approved joint seal in accordance with the manufacturer's instructions,

shall be applied over the complete face.

7.21.2 Where bituminous resurfacing is ramped into an existing bituminous surface and the ramp ends abutting an existing concrete surface, the exposed vertical face of the concrete shall be cleaned thoroughly and either:

- a thin uniform coating of 50 pen or 70 pen hot bitumen; or
- an approved joint seal in accordance with the manufacturer's instructions,

shall be applied over the complete face within 2 hours prior to laying the bituminous surfacing.

7.21.3 The edge of existing concrete surfacing against which a completely new bituminous surfacing is to be laid shall be exposed and thoroughly cleaned to its full depth and for the appropriate length. Unless shown otherwise on the drawings, an expansion joint shall then be formed below the new surfacing by:

- placing a joint filler of non-extruding, heat and rot-proof board, 25 mm thick and of a height equal to the depth of the concrete slab less 100 mm (or, when the total thickness of the new surfacing is greater than this, less the thickness of the surfacing) against the bottom of the exposed concrete face;
- installing an approved joint sealing material in accordance with the manufacturer's instructions to the upper margin of the exposed face; and
- carrying the new surfacing over the top of the joint filler within 2 hours of installing the joint sealing material.

The new surfacing at the junction shall be a minimum of 100 mm thick for a distance of at least 3 m back from the junction, laid in a minimum of two layers.

7.22 Ramps between New Surfacing and Existing Pavements

7.22.1 Junctions to be made with ramps between the new surfacing and existing pavements are shown on the drawings, for which typical details are given in Figures 1 and 2.

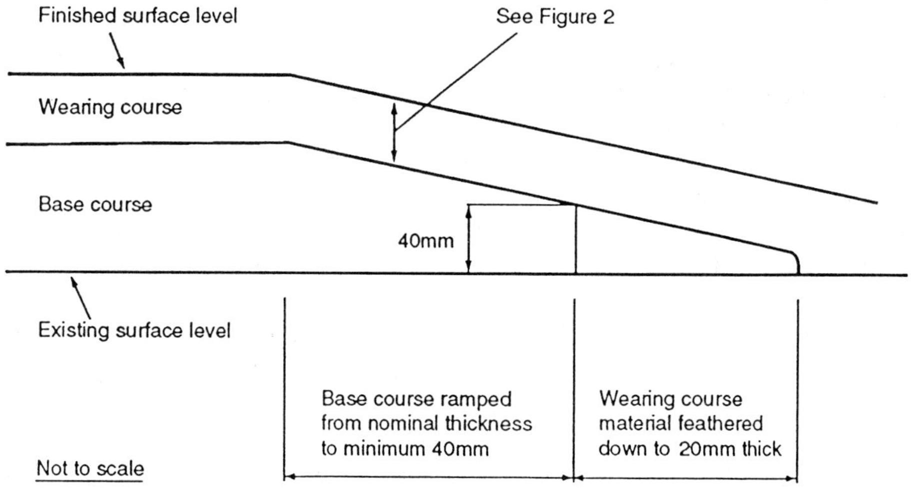

Figure 1 - Ramps between New Surfacing and Existing Pavements

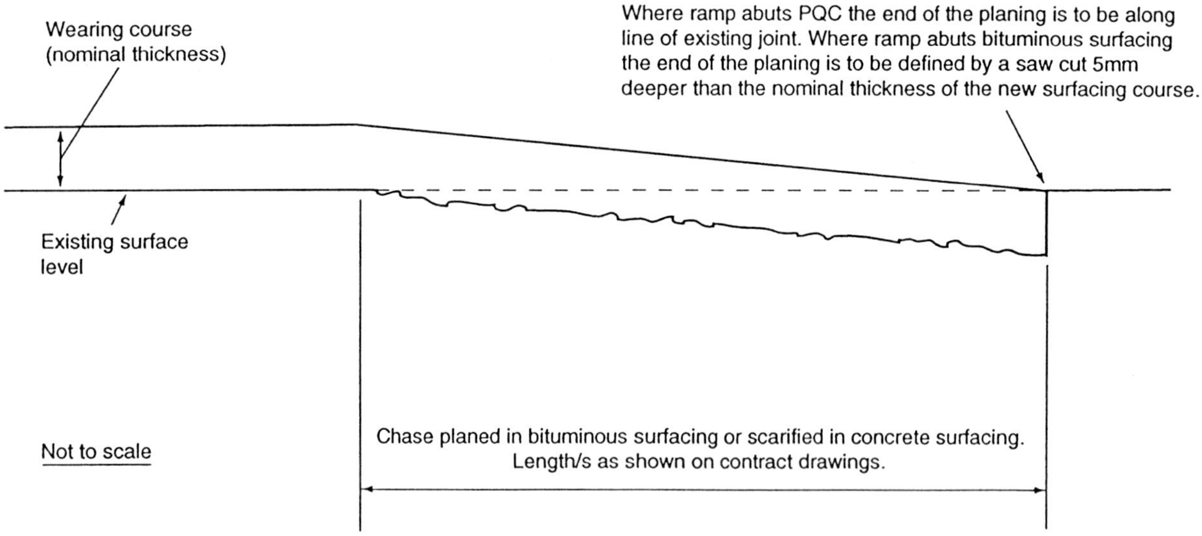

Figure 2 - Ramps between Wearing Course and Existing Pavements

7.22.2 Where appropriate, the new basecourse shall be reduced in thickness at a slope parallel to that of the finished surface to a minimum thickness of 40 mm. It shall be further feathered down at the same slope in hand-laid wearing course or basecourse material using the materials specified in sub-Clause 7.11.2.

7.22.3 The width of these reductions in level shall be as shown on the drawings and shall be formed so the wearing course remains within the tolerances of thickness specified in Clause 7.10 at all points within it. The wearing course shall, however, be permitted to exceed the higher tolerance in the limited area immediately following the end of the basecourse ramping.

7.22.4 Where the existing surface is bituminous, the end of the ramp furthest away from the new surfacing shall be defined by a clean saw cut to the depth shown in Figure 2 before planing is commenced.

7.22.5 Where the existing surface is concrete, the end of the ramp furthest away from the new surfacing shall be along the line of existing joints or grooves. The surface so prepared shall be thoroughly cleared of all loose materials and either:

- a thin uniform coating of 50 pen or 70 pen hot bitumen; or
- an approved joint seal in accordance with the manufacturer's instructions,

shall be applied over the complete face within 2 hours prior to laying the bituminous surfacing.

7.22.6 The exposed vertical edges of the existing surfacing shall be painted with hot bitumen and the new surfacing shall be compacted against it.

7.22.7 In areas where ramping as described above is not considered necessary (i.e. to non-trafficked area), a ramp finished to a step not exceeding 15 mm in the material specified in sub-Clause 7.11.2 is permissible.

7.23 Temporary ramps

NOTE. This is a specimen clause only; the Project/Works Services Manager must agree the details with the Aerodrome Authority/Station staff on a job-specific basis. Further advice for the Project/Works Services manager is given in Clause Z.12 of Appendix Z.

7.23.1 Temporary ramps (e.g. to enable aircraft operations to continue during daylight hours with construction work being carried out at night) shall be constructed in the same manner as specified in Clause 7.22. However, where the Contractor can demonstrate to the satisfaction of the Project/Works Services Manager that he can provide an acceptable and sound surfacing without the use of the saw cut, the saw cutting may be omitted.

7.23.2 Prior to the commencement of each * work, the Contractor shall obtain the approval of the Project/Works Services Manager to the proposed layout of the temporary ramps. At no time shall there be more than one ramp in the length of each Hot Rolled Asphalt or Macadam course and the ramps shall be transverse to the line of aircraft movement or as directed by the Project/Works Services Manager. The ramps shall slope in the same direction unless directed otherwise. On runways, the construction work shall proceed from the primary takeoff/landing end such that temporary ramps slope down in the direction of aircraft movement. If the Contractor proposes to lay any areas of the ramps by hand methods or methods not approved in the trial areas, he shall seek prior approval of the Project/Works Services Manager.

* Project/Works Services Manager to provide the relevant period for the specific job specification; advice is given in Clause Z.12 of Appendix Z.

7.23.3 Temporary ramps shall be arranged so as to avoid covering manholes and aviation ground lighting fixtures whenever practicable. When this is not practicable, the manholes and aviation ground lighting fixtures shall be covered by the temporary ramps.

OR

Temporary ramps shall be arranged so as to avoid covering manholes and aviation ground lighting fixtures wherever practicable. When this is not practicable, ramping, which must be smoothed over as approved and described in sub-Clause 7.23.7, shall also be carried out to all edges of manhole surrounds and aviation ground lighting fixtures.

NOTE. Project/Works Services Manager to select option for specific job specification; advice given in sub-Clause Z.12.2 of Appendix Z.

7.23.4 Ramps on the runway surface within * m of the centre line shall conform with the following requirements:

- The gradient of any ramp shall not exceed ± * in the longitudinal direction;
- The gradient of any ramp shall not exceed ± * in the transverse direction, and there shall be no transverse ramps within * m of the runway centre line;
- The depth of any ramp shall not exceed * mm;
- The spacing of successive ramps shall not be less than * m in the longitudinal direction;
- The profile of the ramps shall be such that it does not create ponding on the surface; and
- All ramps shall be planed into the existing surface to ensure a minimum thickness of * mm of ramping material.

* Project/Works Services Manager to provide values (if clause is required) for specific job specification after agreement with the Aerodrome Authority/Station staff; advice is given in Clause Z.12 of Appendix Z.

7.23.5 Ramps on the taxiway and the extended taxiway to a point * m from the runway centre line shall conform with the following requirements:

- The gradient of any ramp shall not exceed ± * in the longitudinal direction;
- The gradient of any ramp shall not exceed ± * in the transverse direction;
- The profile of the ramps shall be such that it does not create ponding on the surface; and
- All ramps shall be planed into the existing surface to ensure a minimum thickness of * mm of ramping material.

* Project/Works Services Manager to provide values (if clause is required) for specific job specification after agreement with the Aerodrome Authority/Station staff; advice is given in Clause Z.12 of Appendix Z.

7.23.6 Ramps on the runway paved shoulders shall conform with the following requirements:

- The gradient of any ramp shall not exceed ± * in the longitudinal direction; and
- The gradient of any ramp shall not exceed ± * in the transverse direction.

* Project/Works Services Manager to provide values (if clause is required) for specific job specification after agreement with the Aerodrome Authority/Station staff; advice is given in Clause Z.12 of Appendix Z.

7.23.7 All ramps shall be inspected prior to the completion of each * work and are to be modified or made good to the satisfaction of the Project/Works Services Manager where he deems them to be unsatisfactory. There shall be no uncovered tack coat material, oil, petrol, grease or similar contaminants and no steps, ruts, pits or bumps of greater than 3 mm in height/depth. The surface shall be cleaned and swept of all plant, equipment, unsound or ragged materials, debris, dirt and similar, all to the satisfaction of the Project/works Services Manager.

* Project/Works Services Manager to provide the relevant period for the specific job specification; advice is given in Clause Z.12 of Appendix Z.

7.23.8 All or part of the temporary ramps are to be planed out at the commencement of the next period of work to ensure that a minimum of * mm of material can be laid. The planning out of the ramps and laying of the new material is to be carried out in accordance with Clause 7.22.

* Project/Works Services Manager to provide value (if clause is required) for specific job specification after agreement with the Aerodrome Authority/Station staff; advice is given in Clause Z.12 of Appendix Z.

7.24 Finished Levels

7.24.1 The finished surface levels shall conform with the levels, profiles and contours shown on the drawings and the finished levels of the underlying courses are to be such that at no point will the thickness of any overlying courses be less than the thickness specified.

7.24.2 Where the Project/Works Services Manager so directs, deviations from the required levels exceeding 6 mm shall be corrected by replacement with new surfacing at not less than the specified course thickness after removal of the appropriate thickness of the offending area by heating and planing, carried out as specified in Clause 7.3, or by total removal, as detailed in Clause 7.29, when the depth to be removed exceeds 60 mm.

7.25 Surface Accuracy

The surface accuracy of the surfacing material shall be such that the gap between the bottom of a 3 m long test straightedge and the surface of the pavement when the straightedge is placed unsupported on the surface, anywhere in any direction, other than across the crown of a camber or across a drainage channel, does not exceed that specified in the following table:

Surfacing	Maximum Gap under a 3 metre long test straightedge (mm)
Wearing Course	3
Basecourse	10
French drain	10

7.26 Air Voids Content of Designed Hot Rolled Asphalt

7.26.1 After the surfacing has cooled to ambient temperature, at least one 150 mm diameter core shall be extracted as required by Clause 9.6 for the determination of air voids content in accordance with Appendix C (using the same method, A or B, or BS 598: Part 104 as was used for the design). The running mean of 6 cores shall not exceed 5.5 per cent nor be less than 1.0 per cent, and an individual result shall not exceed 6.5 per cent. Where an individual location result exceeds 6.5 per cent, additional cores shall be taken at 10 m intervals on either side of that location and their air voids contents measured to determine the length of non-complying area. Where the running mean of 6 cores exceeds 5.5 per cent, the length of the non-complying area shall be from where the first to where the last of the 6 cores were taken. The non-complying area shall be removed for the full width of the lane. A minimum length of 25 m of defective material shall be removed and replaced with material that shall satisfy the acceptance criteria.

NOTE 1. Reheating a surfacing using an infra-red heater and re-compaction shall not be permitted.

OR

Cores do not have to be taken for the determination of air void content.

NOTE 2. Project/Works Services Manager to select alternative for specific job specification, deleting sub-Clause 7.26.2 if the latter option is chosen; advice given in Clause Z.13 of Appendix Z.

7.26.2 Cores shall be taken so that the whole core is located within 500 mm of the edge of a longitudinal joint for at least one in every three cores sampled.

7.27 Percentage Refusal Density of Basecourse Macadam

7.27.1 After the surfacing has cooled to ambient temperature, at least one 150 mm diameter core shall be extracted as required by Clause 9.6 for the determination of Percentage Refusal Density (PRD) according to BS 598: Part 104. The average PRD of a running mean of 6 cores shall not be less than 93 per cent. Where a length of surfacing does not comply with the average requirement, further cores shall be extracted midway between the locations of the original cores and the new running mean of PRD for 6 cores at the chosen spacing determined. If this running mean does not comply with the PRD requirement, the non-complying area shall be removed for the total lane width. A minimum length of 25 m of defective material shall be removed and replaced with complying material.

NOTE 1. Reheating a surface using an infra-red heater and re-compaction shall not be permitted.

OR

Cores do not have to be taken for the determination of Percentage Refusal Density.

NOTE 2. Project/Works Services Manager to select alternative for specific job specification, deleting sub-Clause 7.27.2 if the latter option is chosen; advice given in Clause Z.13 of Appendix Z.

7.27.2 Cores shall be taken so that the whole core is located within 500 mm of the edge of the pavement for at least one in every three cores sampled.

7.28 Traffic on Finished Surfacing

7.28.1 No trafficking of freshly laid surfacing is permitted until the surfacing has cooled to ambient temperature and no stopping or parking of vehicles is permitted within:

- 24 hours of laying a surfacing in summer conditions; and
- 12 hours of laying a surfacing in winter conditions.

Traffic allowed on finished cold surfacing shall be restricted to the minimum required for the conveyance of mixed materials for the laying of the surfacing immediately adjacent to the area being laid.

7.28.2 If early trafficking of freshly laid material is required, the Contractor shall propose a method of measuring the temperature of the surfacing and validate it during the trials (Section 8).

NOTE. Advice for the Project/Works Services Manager on the assessment of the temperature of freshly laid surfacing is given in Clause Z.14 of Appendix Z.

7.28.3 The Contractor shall be responsible for maintaining the finished surfacing in good and clean condition. He shall make good any defects, damage or defacement which occurs during the Contract by the means, and to the standards, described in this Specification.

7.29 Cutting Out Defective or Old Surfacing

7.29.1 When defective surfacing is cut out, the full depth of the layer shall be removed. The area to be cut out shall extend across the full width of the lane between the longitudinal joints, and shall extend at least 300 mm beyond the defective area into sound material. The area of the lane to be cut out shall be defined by straight saw cuts, at least 40 mm deep or the full thickness of the layer, whichever is the thinner.

7.29.2 The area shall be cut away carefully by pneumatic spade cutters or by approved mechanical scabbling or planing machines. Where spade cutters are used, the cutting tool blades shall overlap the previous cut on each move and each cut shall penetrate the course for its full thickness.

7.29.3 The surface of the exposed material below shall be thoroughly cleared of all loose fragments and cleaned, using high-pressure water and vacuum extraction, or as directed by the Project/Works Services Manager.

7.29.4 Replacement shall not begin until the Project/Works Services Manager has approved the conditions of cleanliness of the base.

7.29.5 When the level of the pavement is to be raised by 20 mm or more, saw cutting shall be omitted.

NOTE. Guidance may be found in Functional Standard 06, "*Guide to Maintenance of Airfield Pavements*" (Defence Works Services 1994).

7.30 Replacement of Defective or Old Surfacing

7.30.1 A tack coat in accordance with Clause 7.9 shall be applied over the exposed base, care being taken to include all corners, angles and irregularities. The exposed vertical faces of the surfacing shall be painted with hot bitumen.

7.30.2 The replacement surfacing shall comply with the specified course thicknesses shown on the drawings and tolerances specified in Clause 7.10, and with the standards detailed in this Specification.

7.31 Filling Core Holes

The walls and base of all holes from which core samples have been cut are to be painted with hot bitumen and filled with the specified surfacing material, well rammed in lifts not exceeding 50 mm.

7.32 Grooving of Hot Rolled Asphalt Wearing Courses

7.32.1 Where the drawings indicate that new or existing surfaces are to be grooved, the grooves shall be formed 4 mm wide and 4 mm deep at 25 mm centres. Grooving operations shall not commence within 24 hours of laying the surfacing and only proceed provided the surface temperature is less than 40°C.

7.32.2 The surface of the wearing course shall be grooved across the pavement at right angles to the pavement edges with grooves which follow across the pavement in a continuous line without break.

7.32.3 The machine for grooving shall be a sawing machine incorporating a minimum of 12 № blades and an automatic guidance system to ensure that the spacings between peripheral grooves on successive passes remain constant. **The Contractor shall state the type and technical details of the machine he proposes to use for grooving at the time of tender.**

7.32.4 At all times during cutting, a high pressure water and vacuum extraction cleaning machine shall be in attendance to pick up the arisings, whether wet or dry, as soon as possible after the grooving machine has passed so that the work area is left in a clean condition free of all loose material and of wet or dry slurry at the end of each shift. Care shall be taken not to wash slurry down the drainage catchpits.

7.32.5 Prior to commencing work, the Contractor shall carry out a trial on an area of disused pavement selected by the Project/Works Services Manager to prove the suitability of the sawing and cleaning equipment. Until approval of the trial is given, grooving work as required by the Contract will not be permitted to start.

8 Trials

8.1 General

Trials shall be carried out on all bituminous mixtures, other than Pervious Macadam drainage material, proposed for use in the Works. For small works, the procedures and requirements for the preliminary and final trials may be combined and/or modified at the discretion of the Project/Works Services Manager.

8.2 Preliminary Trials

8.2.1 For each Asphalt and Macadam, after the design mix or recipe mix has been approved by the Project/Works Services Manager (see Clauses 4.2, 4.3 and 5.2), the Contractor shall mix at least one 20 tonne batch of the material for laying in preliminary trials. At the discretion of the Project/Works Services Manager, this requirement may be modified for regulation or ramping material.

8.2.2 Preliminary trial mixtures, for design or recipe mixtures, shall be made up in the mixers that the Contractor proposes to use at the appropriate mixing temperatures and with the aggregates proportioned from either:

- the various hot bins for a Batch Mixer; or
- the cold feed hoppers for a Continuous Drum Mixer

to produce the required aggregate grading with the binder content at the appropriate design binder content or specified figure for recipe mixtures.

8.2.3 Preliminary trial mixtures for design Hot Rolled Asphalt mixes shall be made up from the laboratory design mixtures determined in accordance with Clause 4.2.

8.2.4 The preliminary trial mixtures shall be laid on disused pavements within the airfield boundary, in locations selected by the Project/Works Services Manager. The materials shall be laid and compacted according to the requirements of BS 594: Part 2, BS 4987: Part 2 and this Specification, according to material type and specified thickness.

8.2.5 The mix shall be transported to the airfield and laid and compacted on a clean firm and level test area, not being part of the works, using delivery, paving and compaction machinery that the Contractor proposes to use during the works. The area shall be tack coated prior to laying. The Contractor shall select paving and compaction operations such that no laying defects such as dragging or surface blemishes result, to the satisfaction of the Project/Works Services Manager. The Project/Works Services Manager may order further 20 tonne loads to be laid until a satisfactory result is achieved.

NOTE 1. Attention needs to be given to the material feed rate via the paver screws to the screed to avoid dragging and segregation.

NOTE 2. For planning purposes, the following equations will assist the Contractor to select paving and rolling rates to achieve the minimum specified number of roller passes before the surfacing has cooled to the minimum temperature for compaction:

Rolling length (m) = average paving speed (m/min) x 8 (min)

Roller passes = (Rolling rate/Paving Rate) x No of Rollers

where: Rolling rate (m^2/min) = Roller width (m) x Roller speed (m/min)
Paver rate (m^2/min) = Paver width (m) x Paver speed (m/min)

8.2.6 The paving and rolling rates, demonstrated to be satisfactory during the laying of the trial mix, shall be agreed with the Project/Works Services Manager and shall be adhered to during the main works.

8.2.7 The Contractor shall submit the results of an analysis, carried out in accordance with BS 598: Part 102, to the Project/Works Services Manager for each load of the trial mix laid to demonstrate compliance with the compositional requirements of Clauses 4.4, 4.5 and 5.2.

8.2.8 If the preliminary trials indicate that the trial mixtures are unsatisfactory for mechanical spreading and compacting, or fail to produce the specified surface accuracy or result in surface blemishes which are unacceptable, the proportion of binder and the grading of the combined aggregates and filler may be slightly modified within the limits of the Specification.

NOTE. If changes outside of the specification limits are required, a new design may be necessary for non-recipe mixtures.

8.2.9 Further preliminary trials shall then be laid to demonstrate that a satisfactory mixture has been achieved. The Project/Works Services Manager will then agree the mixture and appropriate specification limits and authorise the laying of a final trial area as described in Clause 8.3.

8.2.10 After completion of all preliminary trials, the edges of the lanes shall be feathered off in bituminous surfacing as necessary for the requirements of vehicular traffic to the approval of the Project/Works Services Manager.

8.3 Final Trial Areas

8.3.1 Following completion of the preliminary trials described in Clause 8.2, a trial area of surfacing, not less than 60 m or more than 300 m long by two lanes wide, shall be laid for each layer thickness for each mixture. Duplicate areas shall be laid for each additional spreader and for each thickness on which each particular spreading unit will be working. The trial areas shall be laid along the outside edges of the pavements in positions approved by the Project/Works Services Manager.

8.3.2 Each trial area shall contain at least one 60 m continuous length of longitudinal joint and at least 4.5 m of transverse joint.

8.3.3 The Contractor shall submit the results of an analysis, carried out in accordance with BS 598: Part 102, to the Project/Works Services Manager for each lorry load of the trial mix laid to demonstrate compliance with the compositional requirements of Clauses 4.4, 4.5 and 5.2.

8.3.4 When the trial surfacing has cooled to ambient temperature, six 150 mm diameter cores shall be cut using an approved coring machine according to BS 598: Part 100. The cores shall be evenly spaced along the length of the trial surfacing with at least 2 cores being located within 500 mm of the edge of a longitudinal joint. The initial and final cores shall not be taken within 7 m of the ends of the length of surfacing. From these cores, the thickness shall be determined and:

- for Hot Rolled Asphalts, the bulk density and air voids content shall be determined according to the methods in BS 598: Part 104 and Appendix C (using the same method, A or B, or BS 598: Part 104 as was used for the design), respectively; the average air voids content shall not exceed 5.5 per cent nor be less than 1 per cent and an individual result shall not exceed 6.5 per cent.
- for Macadams, the Percentage Refusal Density shall be determined for each core according to BS 598: Part 104; the compaction level shall be deemed acceptable if the average Percentage Refusal Density of the six determinations is not less than 93 per cent.

NOTE 1. There is limited information available on the use of the Percentage Refusal Density Test on wearing course Macadams and experience of the use of this test should be forwarded to the Defence Works Services.

NOTE 2. For very small works and at the discretion of the Project/Works Services Manager, the Air Voids content for Hot Rolled Asphalts and the Percentage Refusal Density Test for Macadams may be omitted.

8.3.5 The Contractor shall demonstrate the effectiveness of his compaction method as defined by the requirements of sub-Clause 8.3.4. The standard of finish, including that of joints, shall comply with the requirements of this Specification and be acceptable to the Project/Works Services Manager as the standard to be achieved in future laying.

8.3.6 When the Project/Works Services Manager has approved the trial area, the Contractor shall confirm in writing the weights, proportions and overall grading of the mixture used in the trial to the Project/Works Services Manager and they shall be used thereafter as targets for future plant mixing with due regard for the tolerances in this Specification.

8.3.7 The approved length of the trial area shall be permanently marked 'Approved' and dated and the approved longitudinal and transverse joints are to be permanently defined.

8.3.8 If the trial area is approved, it shall be allowed to remain. Otherwise, the full length shall be removed as specified in Clause 7.3 and replaced with new material to the required standard at the Contractor's expense.

8.3.9 Until approval has been given, the general laying of bituminous surfacing required by the Contract will not be permitted to start.

8.3.10 The standard of workmanship and finish of all surfacing included in the Contract shall be equal to that of the 'Approved' areas and shall not be changed afterwards without the specific approval of the Project/Works Services Manager.

8.3.11 If for any reason the quality, grading or supply source of any of the aggregates is changed, a new mixture shall be designed and approved on the basis of further trial areas.

8.3.12 Based on the trials, the rollers and rolling method to be adopted for each course shall be agreed with and approved by the Project/Works Services Manager.

8.3.13 No change shall be made afterwards in the mixing and spreading plant or rolling methods without the approval of the Project/Works Services Manager, and then only after new trials have been carried out and approved.

9 Tests

9.1 Test Results

The Contractor shall be responsible for having all testing carried out in accordance with the requirements of this Section and provide the Project/Works Services Manager with a written copy of all results at the first reasonable opportunity but not later than 2 working days after completion of each test. Testing shall be started on specimens within 2 working days of sampling and shall be carried out in an expeditious manner.

9.2 Tests for Initial Approval of Materials

9.2.1 Before mixing starts and as part of the Quality Assurance requirements in Clause 2.4 and sub-Clauses 3.1.2, 3.1.3, 3.4.1, 3.4.3 and 3.5.3, the Contractor (or his materials supplier/s on his behalf) shall have carried out those tests required to determine the properties and grading of the aggregates and filler for comparison with the relevant specification clauses in the following table:

Component or material to be tested	Clause Number	Test	
		Title	Reference
Coarse Aggregate	3.2	Soundness Value Flakiness Index Aggregate Crushing Value Absorption Stripping Polished Stone Value	Appendix A BS 812: Part 105.1 BS 812: Part 110 BS 812: Part 2 Appendix B BS 812: Part 114
Fine Aggregate	3.3	Soundness Value Absorption Stripping *	Appendix A BS 812: Part 2 Appendix B
Filler	3.4	Grading Bulk Density in Toluene	n/a BS 812: Part 2

* Test on parent rock when crushed rock is used

9.2.2 In addition, the Contractor shall submit the appropriate certificates for binder and other constituents, including tack coat and also for tack coat spray bar equipment.

9.3 Test for the Proportioning and Design of Mixtures

As part of the Quality Assurance requirements in Clause 2.4, the Contractor (or his materials supplier on his behalf) shall undertake the series of tests necessary to comply with the relevant specification clauses in the following table:

Component or material to be tested	Clause Number	Test	
		Title	Reference
Designed Hot Rolled Asphalt	4.2	Design Air voids content	BS 598: Part 107 Appendix C
Designed and recipe Hot Rolled Asphalt	4.3, 4.4, 4.5	Analysis	BS 598: Part 102
Pervious Macadam	6.2 6.4	Binder drainage Analysis	Specification for Highway Works BS 598: Part 102
Trials	8.3	Analysis Cores Thickness Bulk density Air voids content Percentage Refusal Density	BS 598: Part 102 BS 598: Part 100 n/a BS 598: Part 104 Appendix C BS 598: Part 104

9.4 Routine Tests on Bulk Supplies Throughout Plant Mixing

9.4.1 The Quality Assurance procedures for the supply of component material and bituminous mixes shall include carrying out tests in order to:

- check on the consistency of bulk supplies;
- compare with the properties and gradings of the samples approved; and
- check on the capability of the dryers to function efficiently with aggregates of variable moisture content.

The tests shall include the following:

Test	Test Method	Representative Sample From	Minimum Frequency	Use For Results
Sieve analysis of aggregates	BS 812: Part 103	Each stockpile	Daily	Comparison with the grading of the initial samples
		Each cold feed hopper	Daily	Computation of the combined grading as an initial check on the production, and for comparison with the approved grading for the mixture being laid

Test	Test Method	Representative Sample From	Minimum Frequency	Use For Results
Moisture content of aggregates at/after mixing	Samples shall be weighed, dried in a ventilated oven at a controlled temperature of 175°C ± 2°C for 24 hours, and then weighed again	After completion of mixing process but without the addition of binder	Weekly	To take measures in order to improve the drying processes if the difference in the weight of the sample before and after oven drying exceeds 0.5 per cent
Sieve analysis of filler	BS 812: Part 103		Daily	Comparison with the results from the initial samples
Bulk density in toluene of filler	BS 812: Part 2		Weekly	

9.4.2 If the result of any test indicates that the bulk deliveries are not of a grading or quality consistent with the approved samples, the Contractor shall, at his own expense, carry out further tests to establish the location and extent to which the materials already stockpiled fail to meet the specified requirements and, if stored on site, shall remove all material condemned by the Project/Works Services Manager for this reason from the aerodrome.

9.4.3 Where the material is mixed off site, any condemned material shall be carefully removed from the stockpiles to be used for the work.

9.5 Routine Tests on Mixtures Throughout Plant Mixing

9.5.1 As part of the Quality Assurance requirements in Clause 2.4, the Contractor (or his materials supplier on his behalf) shall carry out the following tests on mixtures prepared for the works. The procedure shall ensure that the position of plant mixtures from which test samples are taken or specimens made are fully traceable in the finished pavement layer.

Test	Test Method	Sample From	Minimum Frequency	Use For Results
Analysis of the plant mixtures	BS 598: Part 102	After the completion of the mixing process	Every 4 hours but not less than twice a day for each mixer in use	Determination of binder contents and aggregate/filler proportioning and grading - results plotted on graphs in order to show comparison with the grading curve of the mixture approved in accordance with Clause 8.3 and the relevant binder content
Temperature		Lorry	Every load	Check for compliance with Clauses 4.6, 5.3 and 7.8

9.5.2 If either the grading or the binder content of any individual test, out of the total number of tests for the day's production of that mixture, fails to comply with the specified requirements, the additional tests specified in Clause 9.7 shall be carried out.

9.6 Routine Tests during Laying and on Compacted Courses

The Contractor shall undertake the series of tests on bituminous materials incorporated into the works necessary to comply with the relevant specification clauses listed in the following table:

<table>
<tr><th>Test</th><th>Ref. Clause</th><th>Test Method</th><th>Sample From</th><th>Minimum Frequency</th><th>Use For Results</th></tr>
<tr><td>Wind speed and air temperature</td><td>7.2</td><td>n/a</td><td>n/a</td><td>Whenever laying planned</td><td>Assess suitability of weather conditions for laying.</td></tr>
<tr><td>Material temperature</td><td>4.6, 5.3 and 7.8</td><td>BS 598: Part 109</td><td>Paving train</td><td>Each load</td><td>Check on delivery temperature</td></tr>
<tr><td>Surface accuracy</td><td>7.25</td><td>Appendix D</td><td rowspan="2">Previous day's work</td><td rowspan="2">Daily</td><td>Ensure compliance; if more than 2 tests in each group of 20 fail to comply, the additional tests specified in Clause 9.7 shall be carried out</td></tr>
<tr><td>Finished Levels</td><td>7.24</td><td>n/a</td><td>Ensure compliance</td></tr>
<tr><td>Core samples</td><td>7.26 and 7.27</td><td>BS 598: Part 100</td><td rowspan="4">Previous day's work</td><td rowspan="4">Samples from every † laid (or from every 2 hours work, whichever is the more frequent); and at locations agreed with the Project/Works Services Manager</td><td></td></tr>
<tr><td>Course thickness</td><td>7.10</td><td>n/a</td><td>Ensure compliance; if any test fails to comply, the additional tests specified in Clause 9.7 shall be carried out. *</td></tr>
<tr><td>Air voids content</td><td>7.26</td><td>Appendix C</td><td>Ensure compliance; if the test results fail to comply, the additional tests specified in Clause 7.26 shall be carried out.</td></tr>
<tr><td>Percentage refusal density</td><td>7.27</td><td>BS 598: Part 104</td><td>Ensure compliance; if the test results fail to comply, the additional tests specified in Clause 7.27 shall be carried out.</td></tr>
</table>

† Project/Works Services Manager to provide value for specific job specification; advice given in Clause Z.15 of Appendix Z.

* The requirements of Clauses 7.5, 7.6 and 7.9 shall be checked when cores fail in adhesion, and the necessary corrections and adjustments shall be made to eliminate the cause of the failure.

9.7 Additional Tests when Routine Tests on the Mixtures and on the Compacted Courses Fail

9.7.1 The Contractor (or his materials supplier on his behalf) shall carry out the following additional tests when routine tests fail to establish the extent to which material already laid fails to meet the requirements specified:

- for aggregate/filler grading, and binder content;
- for course thickness;
- for air voids content for Hot Rolled Asphalt;
- for percentage refusal density for Macadam; or
- for surface accuracy.

9.7.2 These additional tests (other than for surface accuracy) shall be made on:

- four 300 mm square samples for checking grading and binder content;
- four 150 mm diameter cores for checking course thickness; or
- as specified in Clauses 7.26 and 7.27 for checking air voids content and percentage refusal density, respectively.

The samples or cores shall be cut from the compacted course, at positions selected by the Project/Works Services Manager, within the lane width at a distance of not more than 5 m from the location in the pavement at which the mixture was laid which failed to satisfy the routine test requirements specified.

9.7.3 If any one of the additional test results for aggregate/filler grading, binder content or course thickness also indicate failure to meet the specified requirement, further tests shall be made on 3 more samples or cores. These samples or cores shall be cut at further positions selected by the Project/Works Services Manager, also within the lane width and at a distance of not more than 10 m further along the lane from the location of the previous failure point. Should one of these additional samples or cores also fail to meet the specified requirement, the above process shall be repeated until all samples or cores are satisfactory. The area covered by the failed samples of cores shall be cut out and replaced as detailed in Clauses 7.29 and 7.30.

9.7.4 When the routine tests fail to meet the requirements of Clause 7.25, an additional 20 straightedge tests shall be made over the area between the two routine fail-points, extended by 5 m either side of each along the length of the lane. If three or more of these additional tests also fail to meet the specified requirement, this area of the surfacing shall be condemned. The condemned areas shall be removed and replaced by the Contractor, at his own expense, as specified in Clauses 7.29 and 7.30.

9.7.5 Attempts to correct the surface accuracy with fine bituminous dressings, synthetic resin formulations, surface dressing applications, or emulsion slurry films shall not be allowed.

Appendix A - Magnesium Sulphate Soundness Test

A.1 Scope

This Method shall be followed to determine to determine the soundness of aggregate by subjecting the aggregate to cycles of immersion in a saturated solution of magnesium sulphate followed by oven drying.

A.2 Apparatus and Reagents

Apparatus and reagents as detailed in BS 812: Part 121, Clauses 5 and 6, (except that the balance for coarse aggregate, sub-Clause 5.2, to be accurate to 1 g) together with:

- 37.5 mm, 20.0 mm and 5.0 mm sized square hole perforated plate test sieves and 2.36 mm, 1.18 mm, 600 μm and 300 μm sized woven wire test sieves; the additional test sieves shall comply with BS 410; and
- at least two brass or stainless steel mesh baskets for immersing aggregate specimens for fractions other than 10 to 14 mm with the maximum dimension of the apertures not more than half the maximum aperture of the sieve on which the specimen is retained, but not less than 150 μm.

A.3 Preparation of Test Portions

A.3.1 Bulk samples from each nominal size of aggregate being delivered from each source of supply to be used shall be tested separately and the procedure described hereafter shall be applied to each separate sample.

A.3.2 Prepare two test portions from the bulk samples of each aggregate supplied as in BS 812: Part 121, Clauses 7.1 and 7.2, replacing "*minimum mass of 500 g of the 10.0 mm to 14.0 mm*" in Clause 7.1 by the relevant masses from the tables in Clauses A.4 and A.5.

A.4 Preparation of Fine Aggregate Test Specimens for Each Fraction

A.4.1 The particle size distribution of the test portion shall be determined by the washing and sieving method described in Clause 7 of BS 812; Part 103 using the 10 mm, 5 mm, 2.36 mm, 1.18 mm, 600 μm and 300 μm sieves. The particle size distribution shall be recorded giving the percentage of the mass of the test portion retained between each pair of sieves, together with that passing the 300 μm sieve, to the nearest whole number.

A.4.2 The fraction passing the 300 μm sieve, together with those fractions retained whose proportions are less than 5 per cent by mass of the test portion, shall be discarded; nevertheless, the proportions which they represent shall be taken into account in the calculation of the test result.

A.4.3 One test specimen, of mass in accordance with the table below, shall be taken out of each fraction retained after completion of sub-Clause A.4.2. If there is insufficient material in any of these fractions to provide a test specimen of the required size, the procedure shall be repeated from sub-Clause A.3.2; the particle size distribution recorded shall be that obtained from all the material sieved out.

BS Sieves		Mass of specimen before test (g)
Passing	Retained	
10.0 mm	5.0 mm	300 +10 / -0
5.0 mm	2.36 mm	100 +10 / -0
2.36 mm	1.18 mm	100 +10 / -0
1.18 mm	600 μm	100 +10 / -0
600 μm	300 μm	100 +10 / -0

A.5 Preparation of Coarse Aggregate Test Specimens for Each Fraction

A.5.1 The particle size distribution of the test portion shall be determined by the dry sieving method described in Clause 7.3 of BS 812; Part 103 using the 37.5 mm, 20 mm, 10 mm, 5 mm, 2.36 mm and 1.18 mm sieves. The fractions retained on the 37.5 mm sieve and passing the 1.18 mm sieve shall be discarded and not taken into account in the calculation of the test result. The remainder of the reduced sample shall be considered as the test portion. The particle size distribution shall be recorded giving the percentage of the mass of the test portion retained between each pair of sieves to the nearest whole number.

A.5.2 Those fractions retained whose proportions are less than 5 per cent by mass of the test portion shall be discarded; nevertheless, the proportions which they represent shall be taken into account in the calculation of the test result.

A.5.3 One test specimen, of mass in accordance with the table below, shall be taken out of each fraction retained after completion of sub-Clause A.5.2. If there is insufficient material in any of these fractions to provide a test specimen of the required size, the procedure shall be repeated from sub-Clause A.3.2; the particle size distribution recorded shall be that obtained from all the material sieved out.

BS Sieves		Mass of specimen before test (g)
Passing	Retained	
37.5 mm	20.0 mm	1500 ± 50
20.0 mm	10.0 mm	1000 ± 10
10.0 mm	5.0 mm	300 + 10 / - 0
5.0 mm	2.36 mm	100 + 10 / - 0
2.36 mm	1.18 mm	100 + 10 / - 0

A.6 Procedure

Procedure for each test specimen as in BS 812: Part 121, Clause 8, replacing "*10.0 mm sieve*" in Clause 8.6 by the sieve relevant to the lower size of the aggregate fraction.

A.7 Calculation and Expression of Test Results

A.7.1 Calculate the soundness value of each test specimen as in BS 812: Part 121, Clause 9.1, replacing "*10.0 mm sieve*" by the sieve relevant to the lower size of the aggregate fraction.

A.7.2 Fractions not tested because they represent less than 5 per cent by mass of the test portion shall be assumed to have a soundness value equivalent to:

(a) the mean of the soundness value found by the tests on specimens of the two fractions immediately adjacent to it in size; or

(b) the soundness value found by the test on a specimen of the fraction, either larger or smaller, immediately adjacent to it if only one of these fractions were tested; or

(c) the mean soundness value found by the tests on specimens of the two fractions next but one adjacent to it if both these fractions were tested and the adjacent fractions were not; or

(d) the soundness value found by the test on a specimen of the fraction, either larger or smaller, in this order of priority, most nearly adjacent to it.

A.7.3 For samples of fine aggregate, the material passing the 300 μm sieve shall not be tested but shall be taken as having a soundness value equivalent to that of the specimen passing the 600 μm sieve but retained on the 300 μm sieve.

A.7.4 The soundness value of each test portion of aggregate shall be the sum of the soundness values found for each aggregate fraction times the proportion by mass of that fraction in the test portion.

A.7.5 The magnesium sulphate soundness value (MSSV) for the aggregate shall be the mean of the two results for the test portions to the nearest whole number. The magnesium sulphate soundness value (MSSV) for each fraction of the aggregate shall be the mean of the soundness values for the two results for the test specimens to one decimal place.

NOTE. A suitable worksheet (with two examples, one fine aggregate and one coarse aggregate) is shown below:

Sandy Heath Pit, Coarse Sand, Tested 8-25 August 1994

BS 410 Sieve Size		Grading of Test Portion	Mass of Test Specimen		Soundness Value	Weighted Soundness value
Passing (mm)	Retained (mm)	(per cent of total mass)	Before Test (g)	After Test (g)	(per cent of original mass)	(per cent)
First Test Portion						
10 mm	5 mm	*4.6 ‡*	-	-	*88.1 **	*4.05*
5 mm	2.36 mm	*10.8*	*97.2*	*85.6*	*88.1*	*9.51*
2.36 mm	1.18 mm	*17.0*	*101.8*	*94.2*	*92.5*	*15.75*
1.18 mm	600 μm	*25.2*	*92.9*	*89.0*	*95.8*	*24.14*
600 μm	300 μm	*26.2*	*104.1*	*99.3*	*95.4*	*24.99*
300 μm	-	*16.4*	-	-	*100.0 ★*	*16.40*
	Total	*100*			Total	*94.84*
Second Test Portion						
10 mm	5 mm	*4.4 ‡*	-	-	*88.8 **	*4.08*
5 mm	2.36 mm	*10.9*	*104.1*	*92.4*	*88.8*	*9.59*
2.36 mm	1.18 mm	*17.3*	*106.8*	*98.3*	*92.0*	*15.64*
1.18 mm	600 μm	*25.1*	*101.7*	*96.8*	*95.2*	*23.99*
600 μm	300 μm	*26.1*	*100.3*	*96.1*	*95.8*	*25.10*
300 μm	-	*16.2*	-	-	*100.0 ★*	*16.40*
	Total	*100*			Total	*94.80*
					MSSV	*94.8*

Blackstone Quarry, 6 mm nominal single size, Tested 8-25 August 1994						
BS 410 Sieve Size		Grading of Original Sample (per cent of total mass)	Mass of Test Specimen		Soundness Value (per cent of original mass)	Weighted Soundness value (per cent)
Passing (mm)	Retained (mm)		Before Test (g)	After Test (g)		
First Test Portion						
37.5 mm	20 mm	*0*	-	-	-	*0*
20 mm	10 mm	*0*	-	-	-	*0*
10 mm	5 mm	*25.2*	*303.2*	*278.2*	*91.8*	*23.13*
5 mm	2.36 mm	*69.4*	*104.9*	*98.6*	*94.0*	*65.24*
2.36 mm	1.18 mm	*4.4* ‡	-	-	*94.0* *	*4.14*
1.18 mm	-	*1.2* ‡	-	-	*94.0* †	*1.13*
	Total	*100*			Total	*93.64*
Second Test Portion						
37.5 mm	20 mm	*0*	-	-	-	*0*
20 mm	10 mm	*0*	-	-	-	*0*
10 mm	5 mm	*27.3*	*296.1*	*272.3*	*91.9*	*25.09*
5 mm	2.36 mm	*66.2*	*98.4*	*92.5*	*94.0*	*62.23*
2.36 mm	1.18 mm	*5.1*	*104.1*	*98.2*	*94.3*	*4.81*
1.18 mm	-	*1.4* ‡	-	-	*94.3* #	*1.32*
	Total	*100*			Total	*93.45*
					MSSV	*93.5*

‡ Less than 5 per cent by mass of total sample, no test specimen.

★ No test but mass loss taken as zero for aggregate passing 300 μm sieve, hence Soundness Value of 100 per cent as required by sub-Clause A.7.3.

* Taken as equivalent to that for 5 mm to 2.36 mm size under sub-Clause A.7.2, indent (b).

† Taken as equivalent to that for 5 mm to 2.36 mm size under sub-Clause A.7.2, indent (d).

Taken as equivalent to that for 2.36 mm to 1.18 mm size under sub-Clause A.7.2, indent (b).

A.8 Precision

As in BS 812: Part 121, Clause 10.

A.9 Test Report

As in BS 812: Part 121, Clause 11, together with (d) The magnesium sulphate soundness value and the individual soundness values of the two specimens for each aggregate fraction tested.

Appendix B - Stripping Test

B.1 Scope

This method details the procedure to be followed to determine the resistance of aggregates to stripping.

B.2 Apparatus

B.2.1 Shallow tray(s)

B.2.2 Mixing bowl of 1.5 litres capacity

B.2.3 Means of separately heating the aggregate and the binder to a temperature within the ranges given in sub-Clause B.4.1.

B.2.4 Distilled water.

B.3 Samples

B.3.1 The binder shall comply with Clause 3.5 and shall be a representative sample of the binder to be used in the plant mixtures.

B.3.2 The test shall be carried out on a separate sample of aggregate from each source of supply. The sample shall be taken from the bin or stockpile at the quarry in which the 10.0 - 6.3 mm size predominates. It shall be a representative sample of the rock that is to be used in the plant mixtures.

B.4 Procedure

B.4.1 Each sample of aggregate shall be quartered and sieved out to provide a specimen of 150 particles each passing the 10.0 mm and retained on the 6.3 mm BS sieve. The specimen and the binder shall be heated separately. Under no circumstances shall the binder be heated to a temperature greater than that shown below. The specimen shall be placed in a heated mixing bowl and a quantity of binder equivalent to 4 per cent by mass of the specimen shall be added to it when both specimen and binder are at the following temperatures:

Material	Mixing Temperature
Hot Rolled Asphalt	175°C ± 5°C
Macadam	160°C ± 5°C
Pervious Macadam	125°C ± 5°C

B.4.2 The aggregate and binder shall be hand mixed until coating of the aggregate is complete. If coating is incomplete after 5 minutes of mixing, this procedure shall be repeated with a fresh specimen and an increased proportion of binder. The binder shall be increased by steps of 0.5 per cent by mass of the specimen until a mix giving complete coating of the aggregate is obtained.

B.4.3 The coated specimen shall be placed in one or more trays which have been previously treated with a mixture of equal parts of glycerol and dextrin or similar agent to prevent adhesion of the binder to the tray. Each particle shall be completely separated from adjacent particles.

B.4.4 After standing for one hour, the coated specimen shall be covered with distilled water at 18° - 20°C and maintained at that temperature. After immersion for 48 hours the water shall be decanted and the coated specimen allowed to dry at air temperature.

B.4.5 The dried sample(s) shall then be examined, particle by particle while still in the tray(s).

B.4.6 If the binder has stripped exposing any part of the aggregate from more than six of the total particles tested, the sample shall have failed the test.

B.4.7 To ensure that a fault during test has not contributed to the failure, the test shall be repeated separately on each of a further 3 samples.

B.4.8 If any one of the further tests also fails, by indicating stripping in excess of the limits described in sub-Clause B.4.6, the aggregate will not be acceptable for the particular mixture, and supplies from its source shall be rejected.

Appendix C - Method to Determine the Maximum Density and Air Voids Content

C.1 Scope

C.1.1 This method of test describes the determination of the maximum density of a bituminous mixture and hence the air voids content. It is applicable to all types of bituminous mixture containing penetration grade bitumens or modified binders.

C.1.2 Two methods are described. These are referred to as Method A - the weight in water method, and Method B - the weight in air method.

C.1.3 Samples may be either loose mixtures (e.g. taken from plant or laboratory prepared bulk samples) or they can be prepared from compacted samples taken from the pavement or prepared in the laboratory. The test is suitable for use with both fresh or aged bituminous mixtures.

NOTE. The method is based on a Draft for Development being prepared for the British Standards Institution which, in turn, is based on ASTM D 2041.

C.1.4 If the equipment for test methods A or B is not available, or when more than ten per cent of the aggregate in the total mixture is retained on a 28 mm BS sieve, the air voids can be calculated according to the method in BS 598: Part 104 for each binder content used in the design method using recently determined values for the bulk densities of the various components.

C.1.5 The Contractor shall state that he proposes to use the method in BS 598: Part 104 at the time of tender. Nevertheless, the method described below shall be used in any case of dispute.

NOTE. The methods in this Appendix are preferred because errors in bulk densities do not affect the results.

C.2 Definitions

C.2.1 *The maximum density* of a bituminous mixture is its voidless mass per unit volume.

C.2.2 *Residual pressure* is the pressure above the water in a pycnometer or container after application of a vacuum.

C.3 Summary of Test Methods A and B

A weighed sample of loosened bituminous material is placed in a tared container or pycnometer, covered with water at approximately 25°C, and subjected to a residual pressure or vacuum equivalent to at least 30 mm (4 kPa) of mercury for approximately 15 minutes. After releasing the vacuum, the sample volume is determined by either submerging the container in a water bath and weighing it or by filling the pycnometer with water to a constant volume and weighing it in air. The temperature is measured at the time of weighing and then using the mass and volume measurements, the maximum density of the sample is calculated. Duplicate tests are performed to obtain one result.

C.4 Apparatus

C.4.1 *Ventilated laboratory drying cabinet*, thermostatically controlled to maintain a temperature within the range 80°C to 105°C.

C.4.2 *Sample container*:
Method A - a container having a capacity of not less than 3000 mℓ.
Method B - a volumetrically calibrated pycnometer having a capacity of not less than 3000 mℓ.

C.4.3 *Vacuum desiccator*, or other suitable vacuum vessel, capable of accommodating the pycnometer or container, complete with a safety enclosure to contain the vacuum desiccator and a rubber mat, or other shock absorbent device to prevent impact of the desiccator against a hard surface.

NOTE. The above equipment is only required if the pycnometer or container can not safely withstand the direct application of a vacuum. The safety enclosure is only required if a glass desiccator or vessel is used.

C.4.4 *Vacuum pump or system*, capable of evacuating air from the desiccator, container or pycnometer to a residual pressure of at least 30 mm of mercury (4 kPa).

NOTE. A residual pressure of 30 mm of mercury is equivalent to a vacuum gauge reading of 730 mm at sea-level.

C.4.5 *Residual pressure manometer or calibrated vacuum gauge*, connected directly to the desiccator, pycnometer or container. The manometer or vacuum gauge shall be capable of measuring residual pressure down to at least 30 mm of mercury. If a vacuum gauge is used, there must be an independent means of ensuring the specified vacuum level is achieved.

NOTE. Residual pressure, in mm of mercury, is the difference in the height of mercury in the Torricellian vacuum leg of the manometer and the height of mercury in the other leg of the manometer that is attached to the pycnometer or container.

NOTE. For safety reasons, when a mercury manometer is used, then a trap must be installed in-line, between the manometer and pycnometer or container, to prevent mercury from entering the system. A suitable container must also be positioned under the manometer to catch any spillage of mercury and a small amount of water must be present above the mercury in each leg of the manometer, to prevent vapour losses. Mercury vapour is hazardous and precautions must be taken to prevent and contain any spillage. Fullers Earth or similar can be used to assist with spillages.

C.4.6 *Water traps*, if required, with moisture-absorbing crystals, to prevent moisture entering the vacuum pump. These are installed between the vacuum pump and the desiccator, pycnometer or container.

C.4.7 *Balance*, with sufficient capacity and accurate to 1 g. For method A, the balance must be able to weigh items which are suspended in water.

NOTE. To ensure precision in subsequent calculations, it is necessary to determine the density of samples of uncompacted bituminous mixture to at least four significant figures. A balance with a capacity of at least 10 kg and an accuracy of 1 g is recommended.

C.4.8 *Water bath*, for method A only, of suitable size to accommodate the container, maintained at a temperature of approximately 25°C. The capacity of the bath shall permit free circulation of water around and under the container.

C.4.9 *Sampling apparatus*, in accordance with BS 598: Part 100.

C.4.10 *Thermometer*, capable of reading to an accuracy of at least 0.5°C.

C.5 Bulk Density

Determine the bulk density of each specimen in accordance with the procedure in BS 598: Part 104.

C.6 Test Portion

C.6.1 The test portion shall be three specimens compacted in a Marshall mould for laboratory mixes or cores from the compacted mat for mixed materials. The bulk density of the specimen forming the test portion shall be determined in accordance with BS 598: Part 104 prior to being broken up.

C.6.2 Break up and combine the specimen to form the bulk sample. Obtain a test portion conforming to the requirements of the following table 1 by riffling or quartering the bulk sample in accordance with BS 598: Part 101.

Nominal aggregate size (mm)	Test portion required (g)
37.5	2500 - 4000
28	2000 - 3000
20	1000 - 1500
14	800 - 1200
10	500 - 1000

C.6.3 Place the test portion into the drying cabinet, at a temperature not exceeding 105°C, and dry until it reaches constant weight. If the material is not sufficiently soft to separate manually, warm it in a flat pan in a conventional oven at a temperature not exceeding 120°C, but only until it can be properly handled.

NOTE 1. If the temperature stated is still insufficient to bring the material to a workable condition, it may be increased but shall be kept as low as possible. Increased temperatures shall be reported. Guidance on appropriate softening temperatures is given in BS 598: Part 101, table 1.

NOTE 2. This step can be deleted for test portions prepared from freshly-mixed plant or laboratory mixtures. If required, when aggregates are known to exhibit binder absorption ($> 0.5\%$) freshly mixed samples should be allowed to age for at least 3 hours before test.

C.7 Procedure, General

C.7.1 Separate the test portion so agglomerations of fine coated particles are no larger than 6.3 mm, taking care not to fracture them during the process. If necessary, the test portion may be warmed in the manner described above. If the sample is to be prepared from a core cut from the pavement, ensure the cut and uncoated aggregate faces are re-coated with binder by thoroughly re-mixing the complete core sample after it has softened and is still in a warm condition.

NOTE. It may be found convenient to separate the particles by briskly rubbing the warmed mixture on an old, non-calibrated, sieve.

C.7.2 Allow the test portion to cool to ambient temperature and then proceed using either Method A (Clause C.8) or Method B (Clause C.9).

C.8 Method A - Weight in Water

C.8.1 Weigh the prepared test portion in air (*A*) to 1 g and place in the container.

C.8.2 Add sufficient water, preferably de-aired, at approximately 25°C to cover the test portion by at least 20 mm. The water temperature shall be as close to 25°C as possible and within the range of 22°C to 27°C.

NOTE. If the water temperature is not within the specified range, corrections have to be applied to take account of thermal effects. The correction procedure is cumbersome, compared with measurements within the range specified, therefore, every attempt should be made to ensure appropriate water temperatures are obtained. Corrections, due to thermal effects, within the stated range are negligible and can be ignored. If corrections are required, reference should be made to ASTM D 2041.

C.8.3 Place the cover on the container and remove entrapped air by vigorously shaking it at approximately 2 minute intervals while subjecting the contents to a partial vacuum of at least 30 mm mercury (4 kPa) residual pressure, for 15 ± 1 minutes or such other optimum time as may be determined, commencing from the time the specified vacuum is achieved. If necessary, place the container, without lid, in the vacuum desiccator and de-air as described. A non-foaming wetting agent may be sparingly used.

NOTE 1. If required, establish the optimum vacuum time for a mixture type by reducing the vacuum time until further reduction yields lower bulk densities. The trial reductions shall be performed using a different test portion for each period and at least four time periods shall be used. The minimum time required to remove free air is required, as additional time may cause water to penetrate the binder film and be measured as part of the sample density. Records of the procedure should be kept. In the absence of a specific optimum time, use 15 ± 1 minutes.

NOTE 2. A non-foaming wetting agent such as Aerosol OT in a concentration of 1 mℓ of 10 per cent solution in 100 mℓ of water is recommended. Other materials may also be suitable.

C.8.4 After the vacuum period is complete, fill the container with water, preferably de-aired, at approximately 25°C. Determine its weight in water (*B*), by suspending it completely in the water bath, at approximately 25°C. Determine and record the temperature of the water to 0.5°C and the weight to 1 g.

C.8.5 Empty the container and then re-suspend it completely in the water bath at approximately 25°C to obtain the weight of the empty container in water (*C*) to 1 g. The same operator should repeat the procedure on the second test portion. If the two results differ by more than 11 kg/m^3 repeat the procedure using new test portions.

C.9 Method B - Weight in Air

C.9.1 Calibrate the pycnometer in accordance with the procedure outlined in Clause C.12.

NOTE. A full calibration needs only to be performed infrequently, say every six months. The calibration should be regularly checked by filling the pycnometer with water at a known temperature and, with the lid in place, checking its weight (D) in the calculation below corresponds to the calibrated weight at the same temperature.

C.9.2 Weigh the prepared test portion in air *(A)* to 1 g and place in the container. Add sufficient water, preferably de-aired, at approximately 25°C to cover the test portion by at least

20 mm. The water temperature shall be as close to 25°C as possible and within the range 22°C to 27°C.

NOTE. With some types of pycnometer, it is possible to evacuate the pycnometer before adding water, however, the water should be de-aired in this instance.

C.9.3 Place the cover on the pycnometer and connect it to the vacuum source. Remove entrapped air by vigorously shaking or vibrating the pycnometer at approximately 2 minute intervals while subjecting the contents to a partial vacuum of at least 30 mm mercury (4 kPa) residual pressure for 15 $\pm$ 1 minutes or such other optimum time as may be determined, commencing from the time the specified vacuum is achieved. If necessary, place the pycnometer in the desiccator and de-air as described.

C.9.4 Release the vacuum and remove the cover. Fill the pycnometer with de-aired water at approximately 25°C to the calibration mark or to within 2 - 5 mm of the top. Place the lid or stopper on the pycnometer to expel the air above the water surface.

NOTE. Take care not to apply pressure, when placing the lid or stopper on the pycnometer, such that water is forced out rapidly as this may cause errors. Water should be permitted to trickle out of the spout of the pycnometer in a steady stream, but not such that it breaks contact with the spout.

C.9.5 After drying the outside of the pycnometer, taking care not to draw water out of the spout opening, determine the weight of the pycnometer, sample and water in air (*E*) to 1 g. Remove the lid and measure the temperature of the water *T*°C, to 0.5°C, to ensure it is within the specified range.

C.10 Calculation

C.10.1 For Method A (Weight in water), provided the water temperature is between 22°C and 27°C, calculate the maximum density as follows:

$$\rho_{Max} = \frac{A \times 997.1}{A - (B - C)} \quad kg/m^3$$

where ρ_{Max} = Maximum density (kg/m^3)
A = Weight of dry test portion (g)
B = Weight of test portion and container in water (g)
C = Weight of container in water (g)

NOTE. The density of water at 25°C, at sea level, is 997.1 kg/m^3. The variation in density of water, between 22 and 27°C is approximately 1 kg/m^3 so the use of a more accurate figure than that given above is unnecessary precision.

C.10.2 For Method B (Weight in air), provided the water temperature is between 22°C and 27°C, calculate the maximum density as follows:

$$\rho_{Max} = \frac{A \times 997.1}{(A + D) - E} \quad kg/m^3$$

where ρ_{Max} = Maximum density (kg/m^3)
A = Weight of dry test portion (g)
D = Weight of pycnometer or container filled with water (g)
E = Weight of pycnometer filled with water and test portion (g)

NOTE. See Note 1 to sub-Clause C.10.1 relating to density of water

C.10.3 Calculate the air voids content as follows:

$$Air\ voids\ content = (1 - \frac{\rho}{\rho_{Max}}) \times 100\ per\ cent$$

where ρ = Mean bulk density in accordance with BS 598: Part 104 (kg/m^3)

ρ_{Max} = Maximum density (kg/m^3)

C.11 Precision

Precision data for the test for maximum density is published under ASTM D 2041, where repeatability is stated as 4.0 kg/m^3 and reproducibility as 11 kg/m^3 for single-operator precision. For multi-laboratory precision, the values are repeatability 6.4 kg/m^3 and reproducibility 19 kg/m^3.

C.12 Volumetric Calibration of Pycnometer

C.12.1 The pycnometer used in Method B shall be calibrated prior to use by determining its weight when filled with water at various temperatures, particularly if it is intended to test at temperatures other than those specified above.

C.12.2 Fill the water bath and set to a temperature of 30 to 35°C.

C.12.3 Half-fill the pycnometer with water at about 5°C. Slowly top up with water from the water bath, to about 25 mm from the top of the pycnometer and thoroughly stir the contents.

C.12.4 Place the lid on the pycnometer and, remove entrapped air by vigorously shaking at approximately 2 minute intervals while subjecting the pycnometer to a residual pressure of at least 30 mm mercury (4 kPa) for 15 $\pm$ 1 minutes.

C.12.5 Release the vacuum and remove the lid from the pycnometer. Top-up with water from the water bath and replace the lid, taking care not to apply pressure such that excess water is forced out rapidly as this may cause errors. Water should be permitted to trickle out of the spout of the pycnometer in a steady stream.

C.12.6 After drying the outside of the pycnometer, taking care not to draw water out of the spout opening, determine the weight of the pycnometer and water in air (*D*) to 1 g. Remove the lid and measure the temperature of the water *T*°C to 0.5°C.

C.12.7 Decant about one-third of the water in the pycnometer and again top up with water from the water bath, to about 25 mm from the top and thoroughly stir the contents. Repeat steps 2.3 through 2.5 several times to obtain sufficient data to plot a calibration graph.

C.12.8 Using the data obtained, plot a graph of the weight of the pycnometer filled with water (including the lid) (*D*) against water temperature to obtain a calibration curve for use at various temperatures.

Appendix D - Straightedge Tests

D.1 Scope

This Method shall be followed to determine the surface accuracy of bituminous surfacing layers in this Specification.

D.2 Apparatus

D.2.1 The straightedge for the tests shall be purpose made and 3 m long. It shall have a flat square edge of metal, at least 75 mm wide, along the full length of its base. The straightedge shall be fitted with lifting hand grips or handles.

D.2.2 A calibrated wedge may be used to determine the straightedge clearance. The wedge should have an angle of 5.75° ± 0.05°, and engraved at 10 mm intervals across the incline, starting at the apex, representing clearances increasing in 1 mm intervals up the incline.

D.3 Procedure

D.3.1 The straightedge shall be placed unsupported on the surface, anywhere in any direction, other than across the crown of a camber or across a drainage channel. The location shall be selected by the Project/Works Services Manager or his representative, and the tests shall be carried out in his presence.

D.3.2 Twenty tests shall be made for every 1000 m^2 laid and at least half of these tests shall be across lane joints.

D.3.3 The Contractor shall mark with white paint all areas which fail to comply with the specified requirement.

Appendix E - Guidance on Suitable Temperatures and Wind Speeds for Laying

E.1 In addition to the requirements at Clause 7.2, the following table gives recommended wind speed and air temperature limits for the laying of Hot Rolled Asphalt wearing courses, of Macadam wearing courses and basecourses and of Pervious Macadam drain topping.

Material and Course	Bitumen grade	Thickness (mm)	Max wind speed (km/h) 2 m height	Max wind speed (km/h) 10 m height	Minimum Air Temperature (°C)
Hot Rolled Asphalt Wearing Course	50 pen	40	As in Figure E.1		-
	50 pen	50	{ 40	50	1
			35	45	0
	70 pen	40	As in Figure E.2		-
	70 pen	50	40	50	0
	100 pen	40 - 50	40	50	0
Hot Rolled Asphalt Regulating / Ramping	50 pen	up to 30	See Note 3		
	70 pen	up to 30			
	100 pen	up to 30			
Macadam Basecourse	100 pen	50	{ 40	50	3
			30	40	0
	100 pen	60	40	50	0
DBM 50 basecourse	50 pen	50	See Figure E.3		
	50 pen	60	40	50	0
HDM basecourse	50 pen	50	See Figure E.4		
	50 pen	60	40	50	0
Macadam Wearing Course	200 pen	40	See Figure E.3		
	200 pen	50	40	50	0
	100 pen	40	See Figure E.1		
	100 pen	50	{ 40	50	3
			30	40	0
Macadam Regulating / Ramping	100 pen	up to 50	See Note 3		
Pervious Macadam	200 pen	100	40	50	0

NOTE 1. Where two sets of limiting wind speed and air temperatures are given, laying can proceed provided the weather conditions conform to either set.

NOTE 2. The limiting wind speed and air temperatures relate to a compaction time of 8 minutes for the mid-layer temperature to fall from the specified maximum to minimum compaction temperatures.

NOTE 3. Very thin bituminous layers cool very quickly. At 15 mm thickness less than 2 minutes is theoretically available for compaction within the specified compaction temperature limits. Prompt rolling is essential and it is recommended that this material should only be laid in still conditions.

E.2 Wind speed can be measured by either:

- an anemometer erected at a height of 10 m $\pm$ 0.5 m situated on the airfield; or
- a portable anemometer erected at a height of 2 m $\pm$ 0.1 m situated in close proximity to the laying works.

The anemometer should be fitted with a digital accumulative device. The average wind speed over the previous hour should be used to define the prevailing wind speed.

NOTE 1. If the wind speed is increasing, anemometer readings should be made at 15 minute intervals.

NOTE 2. To aid planning works, weather forecasts may be obtained from the nearest Regional Weather Centre.

E.3 Meterological records for the airfield are available.

OR

No clause.

NOTE. Project/Works Services Manager to select alternative for specific job specification; advice given in Clause Z.16 of Appendix Z.

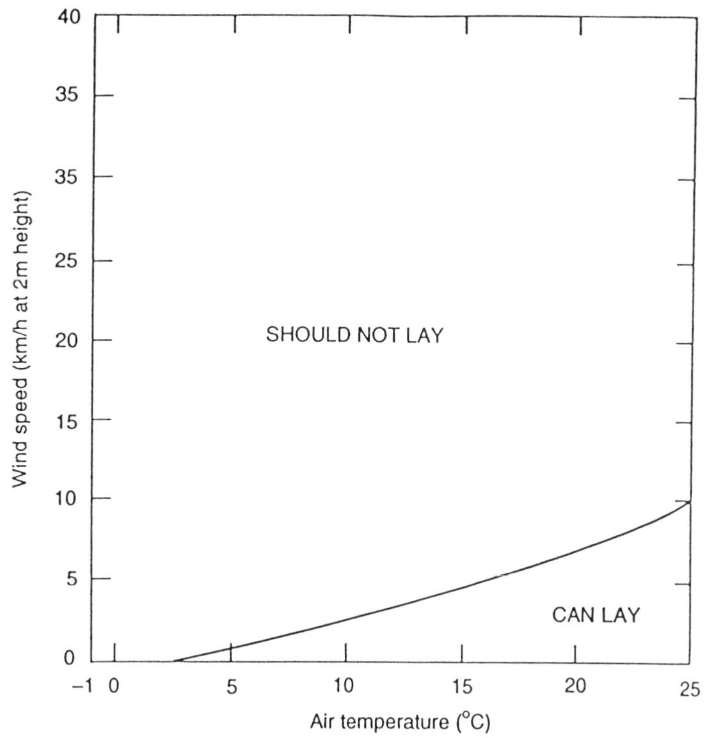

Figure E.1 - Acceptable weather conditions for laying 40 mm thickness of either Hot Rolled Asphalt wearing course with 50 pen bitumen or Macadam wearing course with 100 pen bitumen

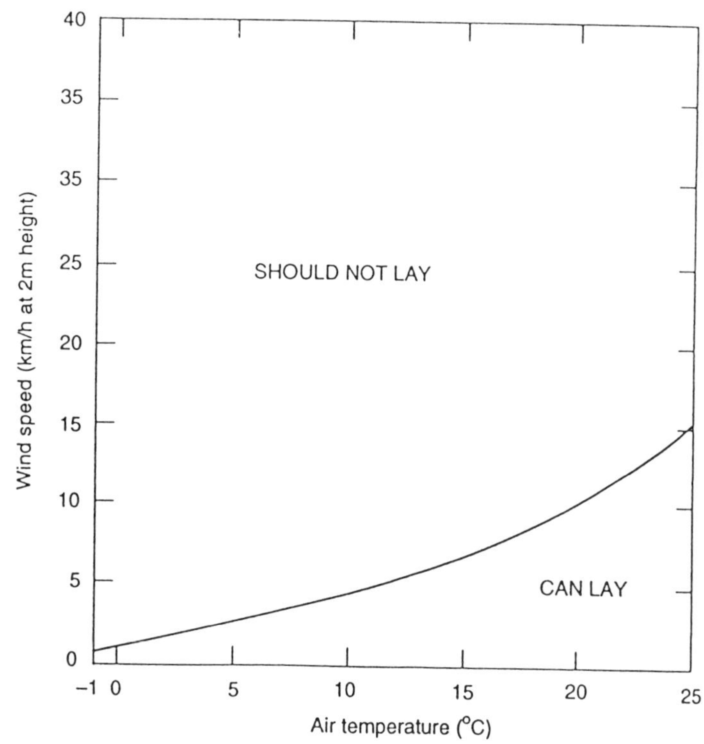

Figure E.2 - Acceptable weather conditions for laying 40 mm thickness of Hot Rolled Asphalt wearing course with 70 pen bitumen

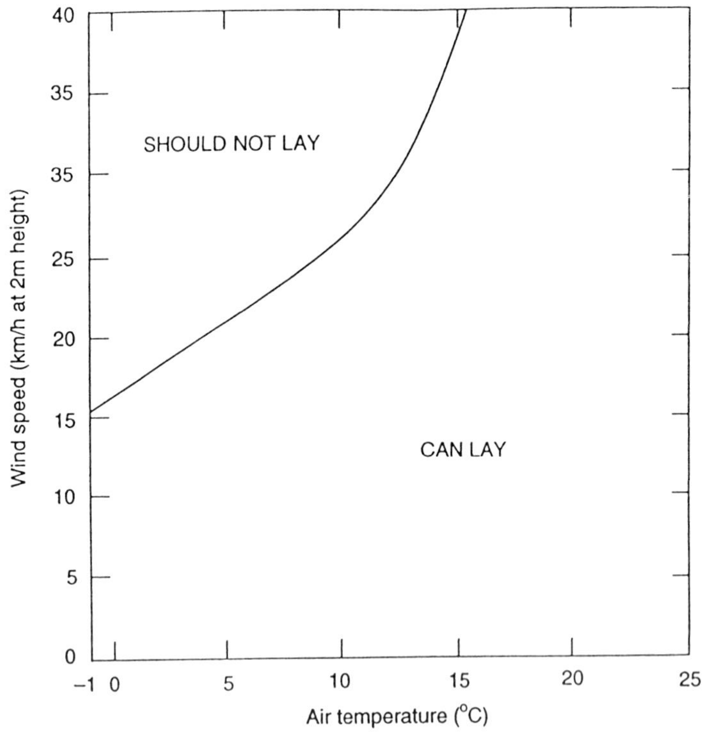

Figure E.3 - Acceptable weather conditions for laying 50 mm thickness of Macadam basecourse with 50 pen bitumen (DBM 50) or 40 mm thickness of Macadam wearing course with 200 pen bitumen

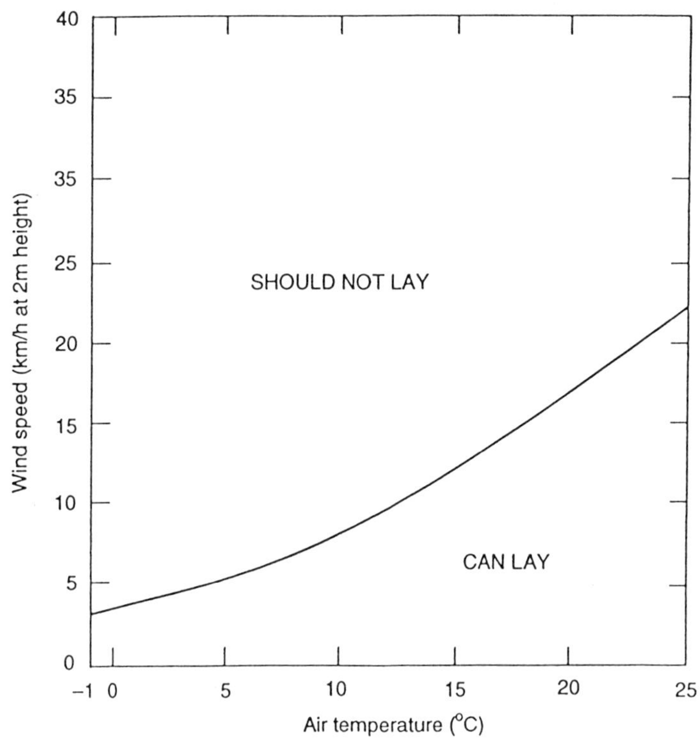

Figure E.4 - Acceptable weather conditions for laying 50 mm thickness of Heavy Duty Macadam basecourse with 50 pen bitumen

Appendix Y - Guidance Notes on Quality Systems for Project/Works Services Managers

Y.1 Introduction

These Guidance Notes are intended to assist Project/Works Services Managers in assessing Suppliers' Quality Assurance (QA) systems for the supply of component and mixed materials as required by Clause 2.4. In particular, guidance is given on:

- how to appraise and evaluate different Quality Systems offered by Suppliers when tendering for jobs (Clauses Y.5 and Y.6); and
- how to monitor work undertaken (Clause Y.7).

Separate clauses are devoted to each aspect.

Y.2 General

Y.2.1 These Guidance Notes are not intended to replace BS/EN/ISO 9000 (formerly BS 5750) and associated documentation.

Y.2.2 The generic term "Client" or "Purchaser" in these Guidance Notes is to include the person or organisation who is acting for, or on behalf of, the Property Manager or the Project Sponsor in the role of either a Project Manager or a Works Services Manager.

Y.2.3 The generic term "Supplier" is used to cover any person or organisation that has, or is tendering for, a Contract with the Client to supply a product or service, and includes those traditionally referred to as the (main) Contractor.

Y.3 Quality Systems

Y.3.1 It is now the accepted practice that all suppliers of goods and services install and maintain a Quality Management System and become registered to a National Accreditation Council for Certification Bodies (NACCB) accredited third party certification scheme as a Supplier of assessed capability.

Y.3.2 It is the general practice that, once a Quality Management Scheme has been set up within an organisation, application for registration to a second or third party certification scheme will be made.

Y.3.3 Second party assessment is carried out by the purchasing organisation; this is very expensive because the resource requirements for systematic and continuous auditing and the management of an assessment scheme are extremely high. Consequently, this form of registration is now in decline and organisation which previously carried out second party assessment, such as British Telecom, British Gas, what used to be the National Coal Board and the MoD, are beginning to insist that their suppliers obtain third party assessment and, hence, pay the costs of quality assurance.

Y.3.4 Certification of a Supplier's Quality Management System by a third party should provide the Purchaser with the confidence that the Supplier is:

- operating and maintaining a fully documented Quality Management System that addresses consistent requirements; and
- operating within the scope of registration.

This avoids the need for the Purchaser to undertake his own structured regime of second party assessments to ascertain the adequacy and focus of the Supplier's Quality Management Systems.

Y.3.5 However, this does not absolve the Project/Works Services Manager of his responsibilities on behalf of the Client to ensure that the Quality System of the Supplier addresses all the requirements and needs. This is because the Quality Standards are interpreted differently by individual organisations.

Y.3.6 In principal, the more independent the assessment and audit regime, the more confident the Purchaser can be as to the value of a Supplier's Quality Management System. Once registered under a certifying bodies scheme, there is still a need to audit by both the Supplier and the third party certification body that procedures and standards are being maintained.

Y.3.7 Assessments and audits can be carried out by:

- The Supplier's management - Under his own audit and monitoring regime;
- The Client (Project/Works Services Manager) - Second party assessment scheme; or
- An independent body - Third party assessment scheme.

Y.4 Processes Covered under the Quality System

Y.4.1 For a Quality Management System to be effective, it must cover all the operations and processes that are relevant to the business conducted by the Supplier.

Y.4.2 Dependent upon the type of work being tendered for, but as a minimum for the purpose of this Standard, the following areas should normally be addressed:

- Procurement, inspection and safe storage of constituent materials;
- Training of plant operatives;
- Setting up on-site mixing plants and the mixing of bituminous materials;
- Off-site supply and mixing of bituminous materials;
- Storage and transportation of bituminous materials prior to use/despatch;
- Inspection and test regimes and records at appropriate stages;
- Sub-contractor/supplier assessment and control;
- Calibration of equipment; and
- Statistical techniques to be used for trend analysis, statistical process control and inspection.

Y.4.3 Quality Management Systems should include provisions for planned and systematic audits, inspections and tests by participating organisations. The Project/Works Services Manager has the responsibility to evaluate and audit the system being operated by the Supplier to ensure adequacy. This should include checking records to substantiate that the procedures are being followed and that the Supplier has evidence that the materials and works are conforming to the specified standard.

Y.4.4 A Quality Plan in accordance with Defence Standard 05-67 should be stipulated in the Contract as a deliverable, but may be included in the *Invitation to Tender* if required. The activities described within the Quality Plan shall be stated unambiguously and concisely so that their intent is clear and that, upon implementation, they can be conducted, assessed, audited, demonstrated, measured or verified.

Y.4.5 The Quality Plan must state or contain definitions as to the levels of quality assurance and control to be applied throughout the Contract, which should include:

- traceability of materials;
- frequency and stages of inspections and tests;
- process controls; and
- records, their retention periods and reviews.

Y.4.6 As quality standards are not extensively defined in relation to the process control, it is the responsibility of the Project/Works Services Manager to ensure that all processes and methods proposed in Quality Plans are clearly defined and understood with regard to how the Supplier will deal with these aspects of the operation, and that all anomalies, shortfalls, errors and omissions are documented and resolved.

Y.5 Assessment of Quality Management Systems

Y.5.1 The assessment of Quality Management System is a logical and progressive multi-stage process that encompasses good management precepts.

Y.5.2 Sub-Clause 6.4.2(c) of BS/EN/ISO 9000 states that a list and records should be maintained of acceptable suppliers. The inference of this is that only suppliers on this list should be chosen. Therefore, any solicited or unsolicited suppliers should, as the first stage in the selection process, be required to complete a supplier questionnaire and must, as a minimum, include the following elements:

- Verification/proof that the Supplier's Quality Management System is registered by a Certification Body accredited by NACCB;
- That the registration certificate is current;
- That the offices/sites the works or services are to be provided from are covered by the registration certificate;
- That the scope of registration is appropriate for the works/services to be provided;
- Experience or references of other users of the Supplier's services;
- The Supplier's past performance, covering experience and results with similar work/projects;
- Financial information;
- Insurance information; and
- Health and Safety information.

Y.5.3 If the responses to the above are satisfactory, an assessment of the Supplier's Quality System can be undertaken.

Y.5.4 The extent of the assessment can range from a visit to the Supplier's premises to overview the Quality Management System in operation on a similar project, to a full formal audit conducted against BS/EN/ISO 9000 and in accordance with BS 7229 by the Project/Works Services Manager's own QA staff.

Y.5.5 It is the responsibility of the Project/Works Services Manager to decide if an assessment is necessary. The decision for, and the scope of, such an assessment should be taken on the basis of the size, complexity, cost and length/duration of the Contract in conjunction with the level of confidence that can be established from other sources.

Y.6 Aspects to Assess Tender Acceptability

Y.6.1 The requirements for the purchasing of goods and services are set down in Clause 4.6 of BS/EN/ISO 9000 and, as a minimum, should be followed. In order to ensure successful procurement, it is a prerequisite that the purchaser (the Project/Works Services Manager) provide a clear definition of requirements in the form of contractual conditions and specifications. This aspect applies equally to Quality Management System requirements.

Y.6.2 All *Invitations to Tender* must contain elements outlining the quality requirements. These should be in the form of asking tenderers to provide:

- proof of registration to the pertinent part of BS/EN/ISO 9000 with an appropriate scope of registration for that particular Contract;
- method statements for all processes to be carried out;
- inspection/test schedules; and
- other information relevant to the Contract.

Y.6.3 When the Supplier returns a tender, his submission must be scrutinised to assess whether his Quality Management System covers all the areas that are relevant to the processes necessary for him to carry out in order that the work is to the required standard. Where not all required elements are covered in the Quality Management System, it may be acceptable for the tenderer to address these areas in his Quality Plan and compile site-specific procedures for unique elements of the Contract.

Y.6.4 The returned tenders must provide precise details against the information requested in the *Invitation to Tender*, which is to include the following:

- The Quality System Standard that will be enforced throughout the duration of the Contract;
- The method and procedures to be used to ensure the positive identification and issue status of specifications, drawings, inspection instructions and other data including the requirements for the approval of operational procedures, equipment, staff, operative training and outputs;
- The method and procedures to be used to ensure the conformance to the Specification by processes, inspection and test criteria; and
- Methods for the procurement of raw materials, services, etc.

Y.6.5 It is the responsibility of the Project/Works Services Manager to select those suppliers which they consider will provide the level of confidence which they require to meet the Specification and fulfil their obligations under the Contract.

Y.6.6 Information, in the form of Inspection Records, Test Certificates and Certificates of Conformity, from the Supplier will not normally be supplied automatically unless particularly requested or a contractual requirement. To make sure that the Supplier understands these requirements, a Quality Plan can be required as part of the tender response. The Quality Plan must be evaluated as part of the tender selection process.

Y.6.7 Where the Supplier proposes that some of the work is carried out by sub-contractors, it does not absolve the Supplier of any of his responsibilities to ensure that the work is carried out to the contracted specification and quality.

Y.6.8 The Supplier is to ensure that, where sub-contractors have their own Quality Management System, it is found by scrutiny to be acceptable and they work to it. Where a sub-contractor does not have his own Quality Management System, the Supplier is to impose his own on the sub-contractor. The Project/Works Services Manager has a responsibility to audit both the Supplier and any sub-contractors to ensure compliance to the tender proposal.

Y.7 Monitoring the Quality Management System and Processes

Y.7.1 Whilst the Supplier may have registration to BS/EN/ISO 9000, it does not necessarily mean that his system is fully focused on the specific requirements of the Contract, nor does any second party scheme run by another purchaser. Monitoring of the system should take place irrespective of whether the Supplier has, or has not, achieved registration.

Y.7.2 The Supplier should have procedures in place for the auditing, monitoring, recording and rectifying of all his activities. The Project/Works Services Manager should ensure, by conducting surveillance audits of the Supplier's system, that:

- these are being carried out;
- the system is effective; and
- the system is focused on the Contract requirements and deliverables.

Y.7.3 Within the Contract, there are requirements for the Supplier to carry out tests on the materials, etc. The Supplier may not have his own test laboratory, in which case he will send samples out to a test house. Any test laboratory, whether part of the Supplier's organisation or an independent test house, conducting the tests should be a NAtional Measurement Accreditation System (NAMAS) accredited test house with an appropriate test schedule.

Y.7.4 Where non-compliancies are found, whether within the system being operated or the goods or services provided, they can be either random instances when the value is outside the specified range or an indication of a trend. If the running mean of the last, say, twenty results has remained reasonably consistent with a standard deviation that also has not fluctuated, then it is likely to be a random instance. Preferably, the running means and standard deviations should be monitored to allow corrective action before non-compliancies occur. All actions taken to deal with non-compliancies are to be documented.

Y.7.5 Rates of sampling and testing must be appropriate to the Contract and stated clearly in the Quality Plan. Where rates are stipulated in the Contract (see Section 9), these will take preference.

Y.7.6 The procedures for sampling and testing Bituminous Materials is to be in accordance with the appropriate parts of the latest editions of relevant British Standards, and also with the latest edition of the appropriate Appendices to this Standard. All samples and testing should be carried out by suitably trained personnel. The results are to be supported by valid Test or Sample Certificates.

Y.7.7 The use of a Quality System should minimise the need for the Project/Works Services Manager to carry out his own tests. Therefore, they can:

- Do nothing because the Supplier is carrying out sufficient inspections and tests, and assessing the results and implications;
- Assess the Inspection and Test Results for the material provided for the Contract to ensure that checks are being made and that the results indicate compliance to the Contract and Quality Plan is being achieved; or
- Conduct a separate inspection and test regime of his own to check for compliance.

Y.8 Records

Y.8.1 The training records of all operatives, sampling and testing personnel are to be maintained by the Supplier and are to be made available for inspection.

Y.8.2 The results of all inspections, tests, etc. for all the Contract should be obtained and retained for record purposes. All documentation (including work-sheets, Inspection and Test Certificates and Certificates of Conformity) that are relevant to the Contract should be:

- available at the place of work (usually the plant or depot) for inspection by the Project/Works Services Manager for the duration of the Contract; and
- handed over to the Project/Works Services Manager on completion of the Contract.

Appendix Z - Guidance Notes on the Preparation of Job Specifications

Z.1 Limitation on the Use of Hot Rolled Asphalt

Z.1.1 Hot Rolled Asphalt may be used for wearing courses in relation to aircraft tyre pressures and frequency of trafficking in accordance with the following table:

Tyre Pressure MPa (psi)	Frequency of Trafficking					
	Low		Medium		High	
	Recipe mix	Designed mix	Recipe mix	Designed mix	Recipe mix	Designed mix
up to 0.7 (100)	Yes	Yes	No (Yes)	Yes	No	Yes
0.7 to 1.4 (100 to 200)	No	Yes	No	Yes	No	Yes
More than 1.4 (200)	No	Yes	No	Yes †	No	Yes †

Yes	Hot Rolled Asphalt is permitted.
No	Hot Rolled Asphalt is not permitted.
(Yes)	Hot Rolled Asphalt is permitted in the cooler, northern regions of the UK only, subject to prior advice being sought from Airfields and Bulk Fuels Group/DWS.
Yes †	Hot Rolled Asphalt is permitted for small works only, subject to prior advice being sought from Airfields and Bulk Fuels Group/DWS.

Z.1.2 The frequency of trafficking is as defined in "*A guide to airfield pavement design and evaluation*" (Property Services Agency 1989) but with the overriding requirements as follows:

Low frequency	Maximum of 50 movements per week by aircraft in the critical tyre pressure range; and
Medium frequency	Maximum of 500 movements per week by aircraft in the critical tyre pressure range.

Z.1.3 Notwithstanding the table in sub-Clause Z.1.1, recipe Hot Rolled Asphalt mixtures may be specified for hand laying small areas.

Z.1.4 The Specification for Hot Rolled Asphalt is for wearing courses only.

Z.1.5 Prior to using this Specification for Hot Rolled Asphalt for runway resurfacing work, the Project/Works Services manager shall obtain advice from the Airfields and Bulk Fuels Group/DWS regarding such matters as grooving and surface texture requirements or the application of coarse textured slurry seal (normally the least favoured option).

Z.2 Limitation on the Use of Macadam

Z.2.1 Macadam may be used for wearing courses and basecourses in relation to aircraft tyre pressures and frequency of trafficking in accordance with the following table:

Tyre Pressure MPa (psi)	Frequency of Trafficking					
	Low		Medium		High	
	Base-course	Wearing Course	Base-course	Wearing Course	Base-course	Wearing Course
up to 0.7 (100)	Yes	Yes	Yes	Yes	Yes	Yes
0.7 to 1.4 (100 to 200)	Yes	No	Yes	No	Yes	No
More than 1.4 (200)	Yes	No	No	No	No	No

Yes Macadam is permitted for that course.
No Macadam is not permitted for that course.

NOTE. Close graded wearing course Macadam as a wearing course may have a low texture depth, and hence poor skid resistance, when new. It is therefore not normally recommended for runways. Consideration should be given to applying a coarse slurry seal when new if surfacing a runway to provide some surface texture, and within 5 years in other situations to seal the surface and arrest binder hardening.

Z.2.2 The frequency of trafficking is as defined in sub-Clause Z.1.2.

Z.3 Use of Pervious Macadam

Pervious Macadam is used as a topping to protect French drains.

Z.4 Soundness of Aggregates

For small works, the Magnesium Sulphate Soundness Test can be omitted provided there is local evidence that the aggregate is sound.

Z.5 Minimum PSV for Wearing Courses

Z.5.1 The minimum requirement for the PSV of coarse aggregate on runway wearing courses should normally be set at 50. For high frequency traffic (as defined in Sub-Clause Z.1.2), a minimum requirement for PSV of 55 should be set.

Z.5.2 The minimum requirement for the PSV of coarse aggregate on taxiway wearing courses should be set at 45. However, for low frequency traffic (as defined in Sub-Clause Z.1.2), the value may be omitted except for fast exit taxiways.

Z.6 Minimum Percentage of Filler Passing 75 μm

The minimum limit on the proportion by mass passing the 75 μm BS sieve should be set at 85 for Hot Rolled Asphalt and 75 for Coated Macadam. On small works, the limit for Hot Rolled Asphalt can be reduced to 75.

Z.7 Hot Rolled Asphalt Type A and B Mixtures

The Type A mixture would be expected to provide a higher stability than the Type B mixtures specified in Clause 4.5.

Z.8 Hot Rolled Asphalt Design Mixtures

Z.8.1 The minimum stability requirements for Hot Rolled Asphalt should be selected according to the following table:

Tyre Pressure MPa (psi)	Minimum Stability (kN)		
	Frequency of Trafficking		
	Low	Medium	High
Up to 0.7 (100)	4.5	6 (4.5)	7.5 (6)
0.7 to 1.4 (100 to 200)	6	7.5 (6)	10
More than 1.4 (200)	7.5	10 ‡	10 ‡

( ) Bracketed values apply in the cooler, northern regions of the UK only, subject to prior advice being sought from Airfields and Bulk Fuels Group/DWS.

‡ Hot Rolled Asphalt should be permitted for small works only subject to advice being sought from Airfields and Bulk Fuels Group/DWS.

Z.8.2 The frequency of trafficking is as defined in Sub-Clause Z.1.2.

Z.9 Bitumen Grades for Use with Macadams

Z.9.1 The grade of bitumen for wearing course and basecourse Macadam may be selected from the following table:

Tyre Pressure MPa (psi)	Frequency of Trafficking					
	Low		Medium		High	
	Base-course	Wearing Course	Base-course	Wearing Course	Base-course	Wearing Course
up to 0.7 (100)	100 pen	100 (200) pen	100 pen	100 (200) pen	100 pen	100 pen
0.7 to 1.4 (100-200)	50 pen ‡	No	50 pen ‡	No	50 pen †	No
More than 1.4 (200)	50 pen †	No	No	No	No	No

( ) Bracketed values apply in the cooler, northern regions of the UK only, subject to prior advice being sought from Airfields and Bulk Fuels Group/DWS.

‡ For use in DBM 50, as given in Clause 6.5 of BS 4987: Part 1 or with increased Filler content in HDM, as given in Clause 6.5 of BS 4987: Part 1.

† For use only with increased filler content in HDM, as given in Clause 6.5 of BS 4987: Part 1.

Z.9.2 In cooler, northern regions of the UK, the use of 200 pen bitumen allows for greater longevity provided adequate structural performance can be achieved. Where 40 mm thick wearing course is specified, the use of 200 pen bitumen also maximises the compaction time.

Z.9.3 The frequency of trafficking is as defined in sub-Clause Z.1.2.

Z.10 Wetting Agents

It is recommended that the option to use Stearine Amine or other approved wetting agent is restricted to small works only.

Z.11 Paving Machines Incorporating Equipment to Apply Tack Coat

The Project/Works services Manager should seek guidance from the Airfields and Bulk Fuels Group/DWS. In the absence of more definitive information, approval would need to be subject to a laying trial and laboratory tests to demonstrate good adhesion, that the tack coat had broken and that moisture had not been trapped.

Z.12 Temporary Ramps

Z.12.1 Phased working to allow aircraft operations to continue, either throughout or at intervals during the construction period (e.g. night working and daytime flying) may necessitate the provision of temporary ramps. Guidance is provided at Appendix 3A of CAP 168, *Licensing of Aerodromes* (Civil Aviation Authority). Clause 7.23 may need to be modified or augmented for specific job specifications and must be subject to prior agreement with the Aerodrome Authority/Station staff.

Z.12.2 The choice of whether to cover over or ramp around manholes and aviation ground lighting fixtures will depend upon, respectively:

- the services located in the manhole and the importance of maintaining immediate access to them; and
- the needs of the Aerodrome to maintain all ground lighting fixtures operational.

Agreement to the choice should be sought with the Aerodrome Authority/Station staff.

Z.12.3 Typical values that can be used in sub-Clauses 7.23.4 to 7.23.6 subject to agreement with the Aerodrome Authority/Station staff are as follows:

	Sub-Clause 7.23.4	Sub-Clause 7.23.5	Sub-Clause 7.23.6
Distance from runway centre line:			
Applicability of sub-Clause	22.5 m	22.5 m	n/a
Minimum for transverse ramps	10.0 m	†	n/a
Maximum longitudinal gradient	1.0 per cent *	2.0 per cent *	2.0 per cent *
Maximum transverse gradient	2.0 per cent	2.5 per cent	2.5 per cent
Minimum spacing of ramps	110 m	†	n/a
Min. thickness of ramp material	40 mm	40 mm	n/a
Maximum depth of ramp	75 mm	†	n/a

* if the existing longitudinal gradient is at, or close to, the maximum permissible in the standard criterion then, depending on the direction of working, it may be necessary to obtain dispensation from the Aerodrome Authority/Station staff to increase this value for temporary ramps.

† Not normally applicable, but otherwise to be agreed with the Aerodrome Authority/Station staff on a job specific basis.

Z.13 Routine Testing

Routine testing of designed Hot Rolled Asphalt for air voids content and of Macadam basecourse Macadams for percentage refusal density throughout the construction period may be omitted for small works only.

Z.14 Temperature of Surfacing

In assessing the temperature of the freshly laid surfacing, consideration should be given to the likelihood of the temperature at depth being greater than that at the surface. If the temperature is required to be taken, a hole should be drilled 40 mm deep, a measurement device with suitable accuracy (± 1°C) inserted and the hole filled with glycerine. The temperature should be determined after the reading has stabilised.

Z.15 Regularity for taking Core Samples

The regularity for taking cores for monitoring should be a minimum of one every 1000 m^2. For small works, the regularity may have to be increased to obtain sufficient results to make an assessment on.

Z.16 Meterological Data

In deciding whether to offer meterological data about the site, the ease of availability of the data and the possible usefulness of the data (in terms of the expected season when the work is to be carried out and the size of the works) need to be considered.

References

Defence Works Services

FS 06		1994	Functional Standard 06, Guide to Airfield Pavement Maintenance

British Standards Institution

BS 12		1991	Specification for Portland cement
BS 410		1986	Specification for test sieves
BS 434			Bitumen road emulsions (anionic and cationic)
	Part 1	1984	Specification for bitumen road emulsions
	Part 2	1984	Code of practice for use of bitumen road emulsions
BS 594			Hot Rolled Asphalt for roads and other paved areas
	Part 1	1992	Specification for constituent materials and Asphalt mixtures
	Part 2	1992	Specification for the transport, laying and compaction of rolled Asphalt
BS 598			Sampling and examination of bituminous mixtures for roads and other paved areas
	Part 100	1987	Methods for sampling and analysis
	Part 102	1989	Analytical test methods
	Part 104	1989	Methods of test for the determination of density and compaction
	Part 107	1990	Method of test for the determination of the composition of design course Rolled Asphalt
	Part 109	1990	Methods for the assessment of the compaction performance of a roller and recommended procedure for the measurement of the temperature of bituminous mixtures
BS 812			Testing aggregates
	Part 2	1975	Methods for the determination of physical properties
	Part 103.1	1985	Methods for determination of particle size distribution: Sieve tests
	Part 105.1	1989	Methods for determination of particle shape: Flakiness index
	Part 110	1990	Method for the determination of aggregate crushing value (ACV)
	Part 114	1989	Method for determination of the polished-stone value
	Part 121	1989	Method for the determination of soundness
BS 890		1972	Specification for building limes
BS 1047		1983	Specification for air-cooled blastfurnace slag aggregate for use in construction
BS 2000			Methods of test for petroleum and its products
	Part 49	1993	Determination of needle penetration of bituminous material
BS 3136	Part 2	1972	Specification for cold emulsion spraying machines for roads
BS 3690			Bitumens for building and civil engineering
	Part 1	1993	Specification for bitumens for roads and other paved areas
BS 4987			Coated Macadam for roads and other paved areas
	Part 1	1993	Specification for constituent materials and for mixtures
	Part 2	1993	Specification for transport, laying and compaction
BS 7229	Parts 1-3	1991	Guide to quality systems auditing
BS/EN/ISO 9000		1994	Quality management and quality assurance standards (replacing BS 5750)
DD 193		1991	Method for recovery of bitumen binders by dichloromethane extraction (rotary film evaporator method)

American Society for Testing and Materials

ASTM D 2041 - 91	1991	Standard test method for theoretical maximum specific gravity and density of bituminous paving mixtures

Civil Aviation Authority

CAP 168		Licensing of Aerodromes

Her Majesty's Stationery Office

Manual of Contract Documents for Highway Works,

Volume 1	1993	Specification for Highway Works
DS 05-67	1980	Defence Standard 05-67, Guidance to Quality Assurance in Design

International Civil Aviation Organization (Quebec)

International Standards and Recommended Practices, Aerodromes: Annex 14 to the Convention on International Civil Aviation,

Volume 1	1990	Aerodrome design and operations
9137 Part 2		Airport Services Manual

North Atlantic Treaty Organisation (Military Agency for Standardization)

Stanag 3634	1992	Runway friction and braking conditions

Property Services Agency

	1989	A guide to airfield pavement design and evaluation

Printed in the United Kingdom for HMSO
Dd. 298034 C9 3/95